PROGRAMMABLE CONTROLLERS

PROGRAMMABLE CONTROLLERS
Workbook and Study Guide

First Edition

Eric A. Bryan
Luis A. Bryan

An IPC Publication
Atlanta

Published by International Programmable Controls, Inc.
 First edition
Printed in the United States of America
89 88 87 86 7 6 5 4 3 2

ISBN 0-915425-01-7

This book was set in Helvetica
Technical revision by Clarence Jones
Edited by Karen Newchurch
Illustrations and art direction by Garon Hart
Manuscript typing and revisions by Diane Buckner

Introduction

This book was designed with a dual purpose in mind: to stand on its own as a study guide and to accompany *Programmable Controllers: Concepts and Applications* as a workbook.

In its role as a study guide, the book enables practicing professionals to refresh their memories and sharpen their skills in PC technology. The generic approach of the text allows immediate application of the information to a variety of industrial automation problems.

Using this publication as a workbook, students and technical personnel encounter a logical progression of information arranged to build upon the topics presented in every preceding unit. The workbook paralles the textbook in its treatment of subject areas, beginning with principles of operations and culminating with selection of the right programmable controller.

Experience has shown that conscientious use of a supplemental book such as this can lead to greater learning and understanding of programmable controllers and their applications. To achieve this goal, the reader must go through every exercise presented in each unit. Section 2 provides all solutions and explanations to the questions presented so that the reader can compare his/her answers.

Whether it is used as part of a structured curriculum or as a brief refresher course for individual PC users or specifiers, *Programmable Controllers: Workbook and Study Guide* will enable its readers to gain valuable insights into current industrial automation technology.

Contents

Section I: STUDY GUIDE AND REVIEW QUESTIONS

Section II: ANSWERS TO REVIEW QUESTIONS

SECTION
--]I[--

STUDY GUIDE
AND
REVIEW QUESTIONS

UNIT --]1[--

INTRODUCTION TO PROGRAMMABLE CONTROLLERS

HISTORICAL BACKGROUND

- The design criteria for the first programmable controller was specified in 1968 by the Hydramatic Division of the General Motors Corporation.
- The primary goal of programmable controllers was to eliminate the high costs associated with inflexible, relay-controlled systems. The specifications required a solid-state system with computer flexibility suited to survival in the industrial environment, easy programming and maintenance by plant engineers and technicians, and reusability.
- Early programmable controllers were used in applications requiring ON/OFF type of control (discrete).
- The first PCs were an improvement over relays since they were easily installed and used less energy and space. Diagnostic indicators incorporated in the PCs aided trouble-shooting.
- In the early 1970s, innovations in microprocessor technology added greater flexibility and intelligence to the PC. Operator interfaces, arithmetic and data manipulation, and computer communications added new dimensions to PC applications.
- Innovations in the mid to late 1970s added even greater flexibility. Improvements included larger memory capacity, remote I/O, analog and position control, and software enhancements. High-speed communication networks started to evolve during these years.
- Present controllers now offer much faster scan times, lower cost, and greater computational capability. Enhancements in hardware include high density I/O systems and intelligent I/O modules.

PRINCIPLES OF OPERATIONS

- A programmable controller is composed of two basic sections: the central processing unit (CPU) and input/output (I/O) interface.
- The CPU is composed of three main parts: the processor, the memory, and the power supply.
- The CPU scans or reads the status of field devices via its I/O interface system. The status of this data is stored in memory and evaluated by the processor as the control or user program is executed. Once the evaluation of the program is completed, the processor updates the status of the outputs according to the evaluation made of the control logic.
- A programming device is used to enter all the necessary instructions to create the control logic program. The most common device used is a cathode ray tube or CRT.

USING PCs

- PCs present an improvement over relays since they eliminate all wiring of interlocking and control relays, thus making easy changes possible.
- A programmable controller is a member of the computer family; however, a PC is made to survive harsh environments and is designed to use a language (ladder) that has been popular for many years by personnel utilizing the old electro-mechanical control relay system.

TYPICAL AREAS OF APPLICATIONS

- The following is a list of the different areas in some industries where PCs are being applied:

Typical Programmable Controller Applications

CHEMICAL/PETROCHEMICAL
Batch Process
Materials Handling
Weighing
Mixing
Finished Product Handling
Water/Waste Treatment
Pipeline Control
Off-Shore Drilling

MANUFACTURING/MACHINING
Energy Demand
Tracer Lathe
Material Conveyors
Assembly Machines
Test Stands
Milling
Grinding
Boring
Cranes
Plating
Welding
Painting
Injection/Blow Molding
Metal Casting
Metal Forming

MINING
Bulk Material Conveyors
Ore Processing
Loading/Unloading
Water/Waste Management

PULP/PAPER/LUMBER
Batch Digesters
Chip Handling
Coating
Wrapping/Stamping
Sorting Winding/Processing
Woodworking
Cut-to-Length

GLASS/FILM
Process
Forming
Finishing
Packaging
Palletizing
Materials Handling
Lehr Control
Cullet Weighing

FOOD/BEVERAGE
Bulk Materials Handling
Brewing
Distilling
Blending
Container Handling
Packaging
Filling
Weighing
Finished Product Handling
Sorting Conveyors
Accumulating Conveyors
Load Forming
Palletizing
Warehouse Storage/Retrieval
Loading/Unloading

METALS
Blast Furnace Control
Continuous Casting
Rolling Mills
Soaking Pit

POWER
Coal Handling
Burner Control
Flue Control
Load Shedding

- PCs are segmented into four major areas:
  - Small (up to 128 I/O)
  - Medium (64 I/O to 1024 I/O)
  - Large (512 I/O to 4096 I/O)
  - Very Large (2048 I/O to 8192 I/O)

- Between each of the four segments, overlapping areas have products that exhibit enhancements of the next (larger) product range.

- PCs with less than 32 I/O are known as microcontrollers and fall in the lower area of the small PC segment.

- The programmable feature of PCs provides the single greatest benefit from the installation of a programmable controller system.

- The following list shows some of the typical benefits that are obtained from the PC features:

Typical Programmable Controller Features/Benefits

Inherent Features	Benefits
Solid-State Components	• High Reliability
Programmable Memory	• Simplifies Changes • Flexible Control
Small Size	• Minimal Space Requirements
Microprocessor Based	• Communication Capability • Higher Level Of Performance • Higher Quality Products • Multi-functional Capability
Software Timers/Counters	• Eliminate Hardware • Easily Changed Presets
Software Control Relays	• Reduce Hardware/Wiring Cost • Reduce Space Requirements
Modular Architecture	• Installation Flexibility • Easily Installed • Hardware Purchase Minimized • Expandability
Variety Of I/O Interface	• Controls Variety Of Devices • Eliminates Customized Control
Remote I/O Stations	• Eliminate Long Wire/Conduit Run
Diagnostic Indicators	• Reduce Trouble-shooting Time • Signal Proper Operation
Modular I/O Interface	• Neat Appearance of Control Panel • Easily Maintained • Easily Wired
Quick I/O Disconnects	• Service w/o Disturbing Wiring
All System Variables Stored in Memory	• Useful Management/Maintenance Data Can Be Output in Report Form

REVIEW QUESTIONS

1-1 In what year and by what company was the first Programmable Controller specified?

1-2 For the purposes of this workbook, PC stands for:

a- Pierre Cardin
b- Printed circuit
c- Personal computer
d- Programmable controller
e- All of the above

1-3 True/False. Even though a more appropriate abbreviation for programmable controllers is PLC (Programmable Logic Controller), in the industrial world they are known as PCs.

1-4 In the early years of their existence, PCs were primarily used for:

a- Position control
b- ON/OFF control
c- Analog control
d- All of the above

1-5 What were some of the intended goals of the first PC design?

1-6 Early innovations of programmable controllers include:

a- Control of PID loops
b- Microprocessor technology
c- Advanced programming languages
d- All of the above

1-7 Innovations that came in the late 1970s included:

a- Large memory capacity
b- Analog I/O
c- Remote I/O
d- All of the above

1-8 True/False. High-speed communication networks added a new dimension to PCs in the sense that they allowed control to be distributed as opposed to centralized.

1-9 True/False. Software enhancements to PCs brought about advanced programming instructions that allowed easy utilizaton of new hardware enhancements.

1-10 Identify the following PC enhancements by specifying whether they are (A)-hardware enhancements or (B)-software enhancements:

____ Fast scan time
____ Functional blocks
____ Data handling
____ Intelligent I/O
____ High-density I/O
____ High-level languages
____ Small, low-cost PCs

1-11 The ____________________ and ____________________ are the two basic sections of a programmable controller.

1-12 Sketch a block diagram of a programmable controller. (Include typical input and output field devices.)

1-13 Sketch a block diagram of the components that form the CPU.

1-14 True/False. The input/output system forms the interfaces through which field devices are connected to the controller.

1-15 The purpose of the I/O interface is to ____________________ the various signals ____________________ from or ____________________ to field devices.

a- Received
b- Connect
c- Sent
d- Condition
e- Control
f- Pre-condition

1-16 A ____________________ device is required to enter the control program into memory.

1-17 A ____________________ is commonly used for program entry, display, and monitoring of the controlled machine or process.

1-18 ____________________ is the process of reading inputs, executing the program, and controlling outputs on a continuous basis.

1-19 What are some distinct differences between PCs and computers?

1-20 Programmable controllers with less than 32 I/O are also known as ____________________.

1-21 True/False. Categorization of programmable controllers is based primarily on I/O count.

1-22 True/False. There are four major categories of PC products. These categories overlap to include products that have enhancements to standard features of the major categories.

1-23 The ______________________ feature provides the single greatest benefit of PCs over hardwired control.

1-24 True/False. In a programmable controller system, there is physical connection between field input devices and output devices.

1-25 True/False. PC systems require as much space in an enclosure as relay systems.

1-26 Explain why the use of remote inputs and outputs is very beneficial in large applications.

1-27 True/False. Programmable controllers can be used to diagnose field device malfunctions.

1-28 Match each of the following features of programmable controllers with the benefits that it offers:

Features	Benefits
a- Solid-state components	____ Minimal space requirements
b- Small size	____ Expandability
c- Software timers/counters	____ High reliability
d- Microprocessor-based	____ Easily changed presets
e- Modular architecture	____ Multi-function capabilities
f- Diagnostic indicators	____ Reduce trouble-shooting

1-29 True/False. Quick disconnects in the I/O interfaces allow module changes without removing the field wiring.

1-30 True/False. Programmable controller systems are utilized to improve system performance and reliability and to produce quality products at reduced costs.

UNIT

--]2[--

NUMBER SYSTEMS AND CODES

- The most commonly found number systems used in programmable controllers are decimal, binary, octal, and hexadecimal (base 10, 2, 8, and 16 respectively).

- Each number system has a base or radix, can be used for counting, can be used to represent quantities or codes, and has a set of symbols.

- The base of a system determines the total number of unique (different) symbols used by that system.

- The decimal equivalent of the largest-valued symbol in any number system is equal to the base minus one.

- The decimal equivalent of any number system can be computed by multiplying the decimal equivalent of each digit symbol (in base b) times the base to the power of its position and then adding the partial multiplications of each previous computation. This procedure is illustrated below.

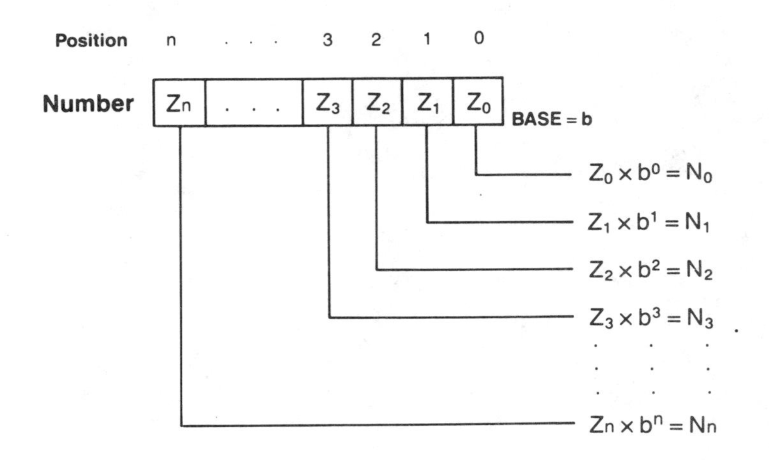

The sum of N_0 through N_0 will the decimal equivalent of the number in base "b."

- The binary number system does not have any symbol greater than the number 1.

- All digital computing devices operate using the binary numbering system.

- The position occupied by a symbol (0 or 1) in binary is known as a bit. A group of eight bits is known as a byte, and two or more bytes form a word or register.

- The octal number system consists of symbols 0,1,2,3,4,5,6,7. There are no 8s or 9s.

- The hexadecimal number system, base 16, consists of 16 symbols including the numbers 0 through 9 and the letters A through F.

- The only operation a digital computing device can perform is addition. Subtraction is accomplished by taking the two's complement of a number (a binary signed negative) and performing an addition.

- The most common binary codes are ASCII, BCD, and Gray codes.

- ASCII (American Standard Code for Information Interchange) is the most common code for alphanumeric representation. This code has 128 characters: 0 through 127 decimal or 0 through 177 octal.

- BCD or Binary Coded Decimal was introduced as a means of representing a decimal number in binary (numbers 0 through 9). This code requires four bits to represent each decimal digit.

- The Gray code is one of a series of cyclic codes known as reflected codes and is suited primarily for position transducers. In this code there is a maximum of one bit change between two consecutive numbers.

- The maximum positive decimal equivalent number that can be represented in a 16 bit binary register is 65,535. If the most significant bit is used as a sign, the maximum and minimum numbers are +32,767 and -32,767.

- The maximum BCD number that can be represented in one register is 9999.

REVIEW QUESTIONS

2-1 True/False. All digital computing devices perform operations in binary.

2-2 Which of the following statements are not true of a number system?

a- Each system has a base.
b- Each system can be used for counting.
c- Each system can be used to represent quantities.
d- Each system has a set of symbols.
e- None of the above.

2-3 True/False. The base of a number system determines the total number of unique symbols used by that system.

2-4 What is the decimal equivalent of the largest-valued symbol in any number system?

2-5 Match the following bases with the appropriate number system:

____	Hexadecimal	a- Base 10
____	Binary	b- Base 8
____	Octal	c- Base 16
____	Decimal	d- Base 2

2-6 Match the largest decimal value equivalent that can be represented by the largest symbol in each of the following bases or number systems.

____	10	a- Decimal
____	2	b- Octal
____	15	c- Binary
____	9	d- Base 11
____	7	e- Hexadecimal
____	1	f- Base 3

2-7 True/False. In any number system, the position of a digit that represents part of the number has a "weight" associated with its value.

2-8 Which is the most commonly used number system for counting or measurement?

2-9 The number fourteen is:

a- 14 in any number system
b- 14 in decimal
c- 14 in octal
d- all of the above

2-10 What would be the equivalent decimal number of any number system of base b as illustrated in Fig. 2-1, and how is it computed?

Position	n	. . .	3	2	1	0	
Number	Z_n	. . .	Z_3	Z_2	Z_1	Z_0	BASE = b

Figure 2-1. Problem 2-10.

2-11 Why is the binary number system used in digital devices such as programmable controllers and computers?

2-12 Convert each of the following binary numbers to its decimal equivalent:

a- 10011011
b- 01100101
c- 11011011
d- 01010101

2-13 In binary numbers, there are only ones and zeros; if you add one plus one the answer would be:

a- One
b- Zero and carry one
c- One and carry zero
d- 2 base 2

2-14 What is a bit in the binary number system?

2-15 What is a nibble, a byte, and a word in a binary system?

2-16 Using the following terms (a through g), fill in the blanks as appropriate.

a- Bit
b- Byte
c- Word
d- Least significant byte
e- Most significant bit
f- Least significant bit
g- Least significant bit of most significant byte

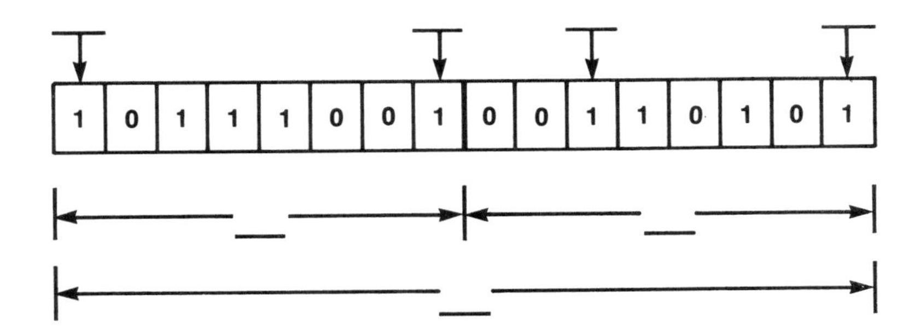

2-17 True/False. A word can be three bytes long.

2-18 True/False. In the octal system, it is not possible to count to an equivalent of eight decimal.

2-19 The largest octal number that can be represented in one byte with all binary ones is ______________________ octal.

2-20 Match the following numbers with their decimal, octal or binary equivalents:

____ 0101	a- 10 decimal
____ 0111	b- 7 decimal
____ 8 decimal	c- 1111 binary
____ 12 octal	d- 5 octal
____ 17 octal	e- 10 octal

2-21 The largest octal number that can be represented in a two byte word is ______________ octal.

2-22 In the hexadecimal number system, the letter F represents a value of 15 decimal. What would be the hexadecimal equivalent of 16 decimal?

2-23 How many bits are required to represent 2B9 hexadecimal in binary?

2-24 What is the largest hexadecimal number that can be represented in a two byte word?

2-25 Convert $IG4_{27}$ to the equivalent number in base 3 given the following information:

Base 27	Decimal	Base 3	Binary
0	0	0	0
1	1	1	1
2	2	2	10
3	3	10	11
4	4	11	100
5	5	12	101
F	15	120	1111
G	16	121	10000
I	18	200	10010

Hint: Notice that $27 = 3^3$

2-26 Convert $A3_{16}$ to its equivalent value in base 4.

2-27 Convert AA_{16} to its equivalent value in base 3.

2-28 Convert the following numbers:

a- 153 octal to binary
b- F35 hexadecimal to octal
c- 28 decimal to binary
d- 35 decimal to octal
e- 101101 binary to decimal
f- 46 octal to decimal
g- 57 octal to hexadecimal

2-29 True/False. One's and two's complement are used to convert a number to its negative equivalent.

2-30 ______________ is the only mathematical operation a digital computing device can perform.

2-31 True/False. The number +20 (decimal) would require six bits to be represented as -20 in its binary equivalent.

2-32 When using the two's complement technique, what would be the largest and smallest numbers represented in a byte?

a- Maximum 01111111 binary; minimum 00000000 binary
b- Maximum +177 octal; minimum 10000001 binary
c- Minimum -177 octal; maximum 11111111 binary
d- Minimum 10000000 binary; maximum 11111111 binary

2-33 Using two's complements, what is the largest and smallest decimal number that can be represented in a 16 bit register?

2-34 Use two's complement to express in binary the answer found in problem 2-33.

2-35 ASCII is a coding system that is used to represent:

a- Alphanumeric and control characters
b- Numbers
c- Characters
d- Control characters

2-36 What does the acronym ASCII stand for?

2-37 The ASCII code generally uses ____ bits and ____ bits when parity is included.

2-38 There are a total of ____ ASCII characters.

a- 127 decimal
b- 177 octal
c- 64 decimal
d- 200 octal

2-39 Representing the decimal number 7 in BCD would require __________ bits.

2-40 What is the decimal equivalent of the 8796 BCD binary pattern?

2-41 How many bits can change state as the number increases in a Gray code scheme?

2-42 What type of field device uses the Gray code?

2-43 True/False. A programmable controller word is also called a register.

2-44 True/False. It is possible to have a four bit word.

2-45 A register can contain __________ BCD digits, each having a maximum value of __________ (BCD).

UNIT
--]3[--

LOGIC CONCEPTS

- There are three basic logic functions used in digital systems: the AND, OR, and NOT functions.

- The binary concept used in logic refers to the fact that many things can be thought of as existing in one of two states. For instance, a light can be ON (1) or OFF (0) and a switch OPEN(0) or CLOSED(1).

- A binary "1" represents the presence of a signal or state in positive logic, while a binary "0" represents the absence of a signal.

- The following are examples of the binary concept utilizing positive and negative logic.

Binary Concept Using Negative Logic

1 (OV)	O (+5V)	Example
Not Operated	Operated	Limit Switch
Not Ringing	Ringing	Bell
Off	On	Light Bulb
Silent	Blowing	Horn
Stopped	Running	Motor
Disengaged	Engaged	Clutch
Open	Closed	Valve

Binary Concept Using Positive Logic

1 (+5V)	O (OV)	Example
Operated	Not Operated	Limit Switch
Ringing	Not Ringing	Bell
On	Off	Light Bulb
Blowing	Silent	Horn
Running	Stopped	Motor
Engaged	Disengaged	Clutch
Closed	Open	Valve

- The AND function is true only if all inputs are true.

- The OR function is true if at least one of the inputs is ON.

- The NOT function is true if the input is not true and vice versa. The NOT function is also called an inverter.

- Boolean algebra was meant to provide a simple way of writing complicated combinations of logical statements that can either be true or false.

- The following table illustrates the logic operations that can be performed using Boolean algebra:

1. Basic Gates. Basic logic gates implement simple logic functions. Each logic function is expressed in terms of a truth table and its Boolean expression.

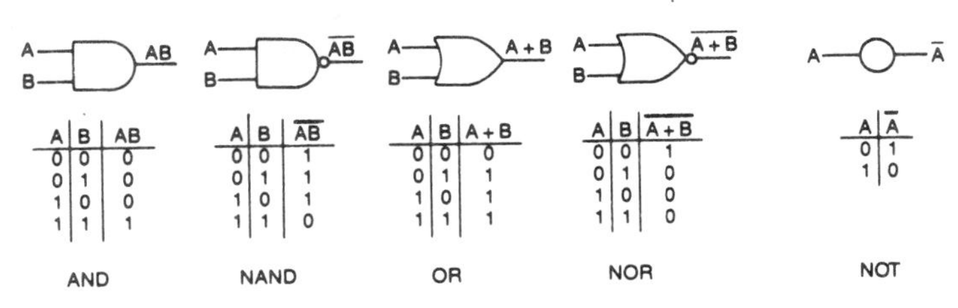

A	B	AB
0	0	0
0	1	0
1	0	0
1	1	1

A	B	$\overline{AB}$
0	0	1
0	1	1
1	0	1
1	1	0

A	B	A + B
0	0	0
0	1	1
1	0	1
1	1	1

A	B	$\overline{A+B}$
0	0	1
0	1	0
1	0	0
1	1	0

A	$\overline{A}$
0	1
1	0

2. Combined Gates. Any combination of control functions can be expressed in Boolean terms using three simple operators: (•), (+), (—).

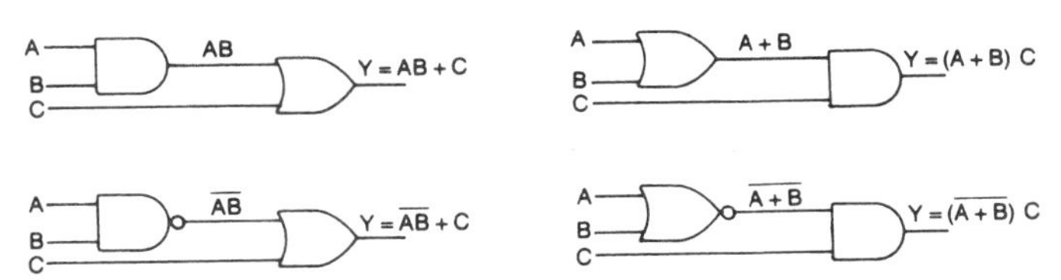

3. Boolean Algebra Rules. Control logic functions can vary from very simple to complex combinations of input variables. However simple or complex the functions may be, they satisfy these basic rules. The rules are a result of simple combination of the basic truth tables and may be applied to simplify logic circuits.

$A + B = B + A$ — *COMMUTATIVE LAWS*
$AB = BA$

$A + (B + C) = (A + B) + C$ — *ASSOCIATIVE LAWS*
$A(BC) = (AB)C$

$A(B + C) = AB + AC$ — *DISTRIBUTIVE LAWS*
$A + BC = (A + B)(A + C)$

$A(A + B) = A + AB = A$ — *LAW OF ABSORPTION*

$\overline{(A + B)} = \overline{A}\,\overline{B}$ — *DE MORGAN'S LAWS*
$\overline{(AB)} = \overline{A} + \overline{B}$

$\overline{\overline{A}} = A\,,\ \overline{1} = 0\,,\ \overline{0} = 1$

$A + \overline{A}B = A + B$

$AB + AC + B\overline{C} = AC + B\overline{C}$

4. Order of Operations and Grouping Signs. The order in which Boolean operations (AND,OR,NOT) of an expression is performed are important. This order will affect the resulting logic value of the expression. Consider the three input signals A,B,C. Combining them in the expression Y = A + B•C can result in misoperation of the output device(Y), depending on the order in which the operations are performed. Performing the OR operation prior to the AND operation is written (A + B)•C, and performing the AND operation prior to the OR is written A + (B•C). The result of these two expressions is not the same.

The order of priority in Boolean expressions is NOT (inversion) first, AND second, and OR last, unless otherwise indicated by grouping signs such as parentheses, brackets, braces, or the vinculum. According to these rules, the previous expression A + B•C, without any grouping signs, will always be evaluated only as A + (B•C). With the parentheses, it is obvious that B is

Continued

ANDed with C prior to ORing the result with A. Knowing the order of evaluation then makes it possible to write the expression simply as A+BC, without fear of misoperation. As a matter of convention, the AND operator is usually omitted in Boolean expressions.

When working with Boolean logic expressions, misuse of grouping signs is a common occurrence. However, if signs occur in pairs, they do not generally cause problems if they have been properly placed according to the desired logic. Enclosing within parentheses two variables that are to be ANDed is not necessary, since the AND operator would normally be performed first. If, however, B is to be ORed with C prior to ANDing it with A, then B+C must be placed within parentheses.

When using grouping signs to insure proper order of evaluation of an expression, parentheses ( ) are used first. If additional signs are required, brackets , and finally braces [ ] are used. An illustration of the use of grouping signs is shown here.

$$Y1 = Y2 + Y5 \quad [X1(X2 + X3)] + [Y4\,Y3 + X2(X5 + X6)]$$

5. Application of DeMorgan's Laws. DeMorgan's Laws are frequently used to simplify inverted logic expressions or simply to convert an expression into a useable form. See Appendix B for other conversions using DeMorgan's Laws.

According to DeMorgan's Laws:

$$\overline{AB} = \overline{A} + \overline{B} \qquad \text{and} \qquad \overline{A + B} = \overline{A}\,\overline{B}$$

A, B → $Y = \overline{AB}$ → $\overline{A}$, $\overline{B}$ → $Y = \overline{A} + \overline{B}$ A, B → $Y = \overline{A+B}$ → $\overline{A}$, $\overline{B}$ → $Y = \overline{A}\,\overline{B}$

- Hardwired logic refers to logic control functions (timing, sequencing, and control) that are determined by the way devices are interconnected.

- Relay logic implemented in PCs is based on the three basic logic functions (AND,OR,NOT), which are used singly or in combinations to form instructions that will determine if a device is to be switched ON or OFF.

- Ladder diagrams, also called contact symbology, are relay-equivalent contact symbols (i.e., normally open and normally closed contacts and coils) used in the programming of control logic used in PCs.

- Symbols in ladder diagrams can be in series, parallel, or a combination of both.

REVIEW QUESTIONS

3-1 The binary concept is used in digital systems because:

a- A light can be ON or OFF
b- Voltage levels can be low or high
c- The two-state concept conditions
d- None of the above

3-2 Which logic type is more conventional?

a- Positive logic
b- Negative logic
c- Depends on the application
d- None of the above

3-3 The logic operations used by PCs are:

a- AND
b- OR
c- NOT
d- All of the above

3-4 If you are to relate the AND function with a mathematical operation, which of the following would best describe the function?

a- Division
b- Multiplication
c- Addition
d- None of the above

3-5 True/False. Logic gates and Boolean gates do not represent the same functions.

3-6 True/False. Ladder diagrams are also called contact symbology.

3-7 True/False. In most PCs, a rung can contain more than one output.

3-8 The NOT function is also called:

a- False-to-true converter
b- Changer of states
c- An inverter
d- All of the above

3-9 Using the following elements, describe the guidelines for translating each from hardwired logic to programmed logic:

a- The normally-open contact
b- The normally-closed contact
c- An output
d- Repeated use of relay contacts

3-10 A NOT symbol is used when a logic "1" must ____________________ some device.

a- Activate
b- Deactivate
c- Change
d- None of the above

3-11 Assuming positive logic, translate the following hypothetical examples into Boolean gates (use AND, OR, NOT gates).

A DC-8 plane needs the following requirements to take off:

a- All four engines must be at full power.
b- Three out of four flight attendants must be present
c- There should be no other planes landing or taking off
d- They must receive the go ahead from ground control

3-12 True/False. The binary concept is a new concept that has been applied in the past fifteen years.

3-13 The AND function, implemented using contacts, will mean contacts:

a- In parallel
b- In series
c- a and b
d- Will depend strictly on the application

3-14 Define the term rung.

3-15 Express the following equation as a ladder diagram rung:

$y = [(\bar{a} + \bar{b}) \cdot c] + [(d \cdot e) + f]$

3-16 Choose all of the following statements that apply to a single coil:

a- It could be used as an internal output
b- Its contacts could be used as a normally closed input in a rung
c- It could be used to drive several real outputs in more than one rung
d- All of the above

3-17 Using a,b,c,or d, indicate the type of circuit for each logic sequence shown in Fig.3-1.

a- Series/parallel circuit
b- Parallel/series circuit
c- Series circuit
d- Parallel circuit

3-18 True/False. Boolean algebra was developed during the early stages of the computer era.

3-19 In a ladder diagram control program, the output coil is always displayed:

a- At the far left
b- At the far right
c- In the center
d- Depends on the control program

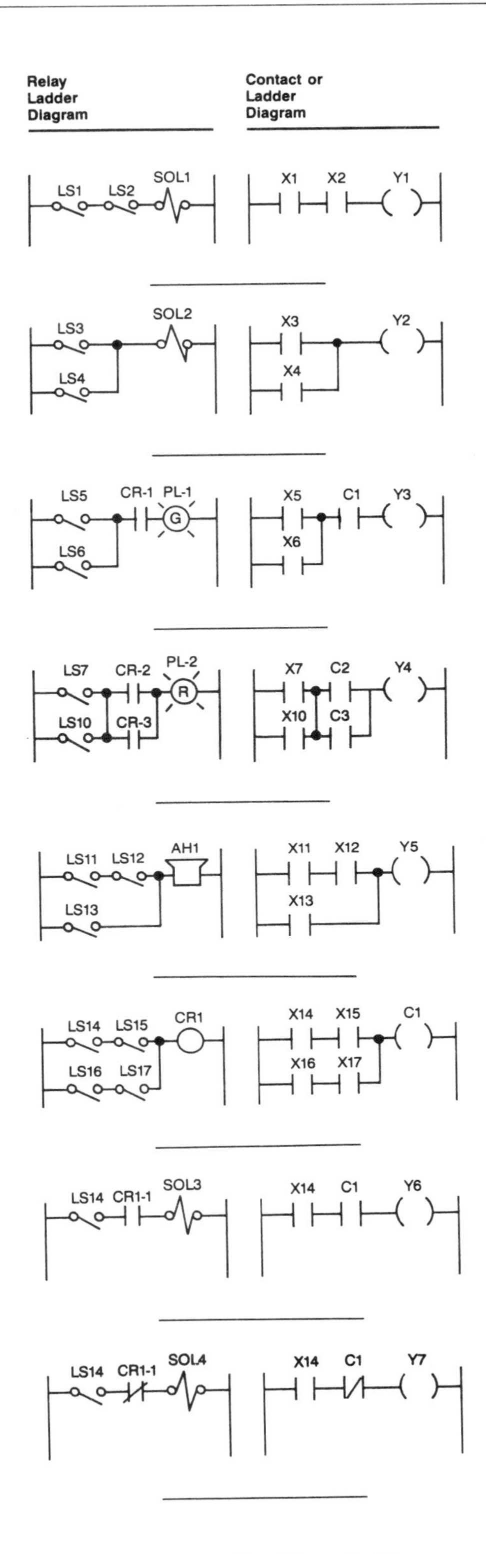

Figure 3-1. Problem 3-17.

3-20 True/False. The interconnection of devices in a programmable controller-based system is very important since it will determine how logic control functions are executed.

3-21 Draw the symbol for the NOR gate and show the truth table.

3-22 The ____________ output or coil is used to deactivate an output device when any left-to-right path of input conditions are true.

3-23 True/False. The allowance of repeated use of contacts in the control program is dependent on the controller.

3-24 The NOT function implemented in contact symbology is:

a- A normally closed contact
b- A normally open contact
c- An internal output
d- None of the above

3-25 True/False. The maximum number of contacts in ladder diagram logic is a matrix of 7 x 7.

3-26 ________________________ are used to provide additional contacts for interlocking logic.

3-27 True/False. An address for a given input/output can only be used once in a ladder program unless additional wiring is done.

3-28 Draw the symbol for the AND gate using three inputs. Show the truth table.

3-29 Implement the following Boolean equation using logic gates; then minimize the equation using Boolean algebra laws and implement the simplified equation.

$$[(a + ab) + (\bar{a} + \bar{b})]\,[\bar{b} + (a + \bar{a}b)] = Y$$

UNIT

--]4[--

THE CENTRAL PROCESSING UNIT

- The central processing unit (CPU) is composed of these main sections: the processor, the memory systems, and the system power supply.
- The term CPU is often used interchangeably with processor; however, the CPU term encompasses all the necessary elements that form the intelligence of the system.
- The processor is continually interacting with the system memory to interpret and execute the application program that controls the machine or process. The power supply provides all the necessary voltage levels to insure proper operation of the processor and memory components.

PROCESSOR

- The intelligence of today's programmable controllers is formed by small microprocessors (micro) with tremendous computing and control capability.
- The principal function of the processor is to command and govern all the activities of the entire system.
- The processor performs the system supervision by interpreting and executing a collection of system programs known as the executive.
- Multi-processing is an approach that divides the total system tasks among several microprocessors, thus sharing the control and processing responsibility.
- Microprocessors operate on registers or words of different lengths (4, 8, or 16 bits). The longer the word length, the faster a micro operates.
- The process of reading all inputs, executing the control program, and updating all outputs is known as scan.
- The scan time is generally specified as a function of user memory used (usually in terms of K units).
- Immediate instructions interrupt the normal program scan and perform an I/O update to read inputs or write outputs.
- If an input signal changes states twice in one scan, the PC will never be able to "see" the signal; this situation occurs if the change in signal is faster than the scan time.
- Parity check is also called vertical redundancy check (VRC). This error-checking method checks for an even or odd number of 1s, depending on whether the check is for even parity or odd parity.

- The parity bit is used to make each byte or word have an odd or even number of 1s.
- Parity error-checking is a single-error detection method with no correction capability.
- The checksum error detection method provides a means of checking a block of data. The last word of the block, the block check character (BCC), contains a pattern that describes the bit organization of the whole block.
- The most common checksum error detection methods include cyclic redundancy check (CRC), longitudinal redundancy check (LRC), and the cyclic exclusive-OR checksum.
- The most common error detection and correction code is the Hamming Code, which can detect two or more bit errors and correct one-bit errors.
- Typical CPU diagnostics include memory, processor, battery, and power supply.

THE MEMORY SYSTEM

- The memory of a PC is composed of the system memory and application memory sections. The system memory is composed of the executive and scratch pad sections. The application memory is formed by the data table, which includes the input/output table, internal bits storage area, and the user program area, which stores the control program.
- There are two types of memories: volatile and non-volatile. Volatile memory loses its contents when power is removed; non-volatile memory retains its contents when power is removed.
- The system memory, which includes the executive, is stored in non-volatile memory. The user memory is generally stored in volatile memory with battery back-up.
- A new type of memory technology, known as non-volatile random access memory (NOVRAM), combines the non-volatility of the EEPROM and the speed performance of the RAM.
- The most basic unit of the memory structure is a cell or bit. Each cell or bit stores information in binary in the form of voltage (true,1) or novoltage (false,0).
- The abbreviation K represents 1024 locations or words when referring to memory capacity. A 2K represents 2048 locations, a 4K, 4096, and so on.
- The amount of memory required by each instruction in a PC is known as memory utilization.
- The input/output table is a section in memory which changes constantly and reflects the changes of inputs (input map) and the updates of outputs according to the control program.
- Internal storage bits are used as internal coils for interlocking of the control program.

- Storage registers are classified in three types: input registers, holding registers, and output registers. These registers can hold BCD or binary values.
- The system power supply plays an important role in the total PC system operation. It not only supplies all the necessary voltages to the system but also serves as first defense of system reliability and integrity by monitoring proper levels of the supplied voltages.
- A power down sequence is initiated by a command from the power supply to the processor when the line voltage exceeds the upper or lower limits for a specified duration (usually 1-3 AC cycles).
- Constant voltage transformers should be incorporated in the system when line voltages vary beyond the limits of the power supplies too often.
- Isolation transformers should be used in applications where high noise is generated and infiltrated into the power supply line.

REVIEW QUESTIONS

4-1 The term CPU (Central Processing Unit) is often used interchangeably with ____________________.

4-2 True/False. The term CPU encompasses the elements that form the intelligence of the system.

4-3 The primary component of today's programmable controller's CPU is the ____________________.

4-4 The processor governs all systems activities by interpreting and executing a collection of system programs known as the ____________________.

4-5 True/False. The processor has no interaction with the PC system communication.

4-6 An approach used to divide the total system duties among several processors that share the control and processing responsibilities is known as ____________________.

4-7 I/O interfaces that are microprocessor-based allow ____________________ control to take place outside of the main processor.

4-8 True/False. The speed at which a microprocessor can solve a program is also a function of word length.

4-9 True/False. The scan time is based on the total amount of memory of the system.

4-10 The use of remote I/O ______________________ the scan time as a result of the remote subsystem transmission.

4-11 The total scan time of a system includes the ______________________ scan time plus the ______________________ time.

4-12 Explain how immediate instructions operate during a normal scan time and make a functional sketch.

4-13 If a PC has a total scan time of 7 milliseconds and has to monitor a signal that ______________________, then the controller will never "see" the signal.

- a- Changes state once in 14 msec
- b- Is fast
- c- Changes state twice in 5 msec
- d- Is constantly changing

4-14 The communication between the main CPU and the remote I/O takes place in:

- a- Parallel format
- b- Serial binary format
- c- The CPU
- d- The remote driver

4-15 ______________________ techniques are normally incorporated in the communication between the CPU and a remote subsystem to confirm the validity of the data transmitted.

4-16 The parity check is sometimes called ______________________.

4-17 Parity error-checking examines the transmitted data for an ______________________ or ______________________ number of 1s.

4-18 ______________________ is the extra bit that is incorporated in the data transmission using parity check.

4-19 If the processor is transmitting the 7 bit ASCII character "E" (105_8) to a peripheral device, and odd parity is required, what must the parity bit be? In the same transmission, if even parity is required, what must the parity bit be?

4-20 Parity error checking can detect:

- a- Multiple errors
- b- Single errors and correct them
- c- Single errors
- d- Double errors and correct one

4-21 True/False. If during the transmission (with even parity) of ASCII character "A" (1000001), an ASCII "B" (1000010) is received, then an error is detected.

4-22 True/False. Checksum is an error detection technique that is used in transmission of multiple words.

4-23 The last word in a checksum computation is known as:

a- End check character
b- Character check
c- Block check character
d- Block character check

4-24 Which of the following is not a checksum method?

a- Cyclic redundancy check (CRC)
b- Horizontal redundancy check (HRC)
c- Longitudinal redundancy check (LRC)
d- Cyclic exclusive-OR checksum

4-25 True/False. Most error detection methods are performed by a software routine in the executive program.

4-26 The ______________________ is the most commonly used error detection and correction code.

4-27 Error-correcting codes offer the advantage of being able to detect ______________________ bit errors and can correct ______________________ bit errors.

4-28 Name at least four CPU diagnostics that are performed by the controller and possibly indicated on the front of the CPU.

4-29 Relay contacts available in the programmable controller that act as fault contacts operate in a ______________________ timer fashion.

4-30 Match the following sections of a typical PC memory system with its use:

____ Executive	a- Control program area
____ Scratch pad	b- Store constants area
____ Application memory	c- System software area
____ Data table	d- Interim calculations area

4-31 True/False. The executive software in most programmable controllers is backed up by a battery.

4-32 What are the two categories of memory that are found in PCs?

4-33 A ________________ memory will lose its programmed contents if operating power is lost.

4-34 A ________________ memory will retain its programmed contents if operating power is lost.

4-35 ________________ is designed to permanently store a program that cannot be altered under any circumstances.

4-36 ________________ is often referred to as read/write memory and also can lose the program if power is lost.

4-37 ________________ is the most commonly used application memory and is generally backed up by a battery.

4-38 ________________ is designed to be reprogrammed after being erased with ultra-violet light.

4-39 ________________ is a memory type that can be erased electrically, and its contents are preserved during a power down.

4-40 Executive programs are generally stored in ________________ memory.

4-41 ________________ is a RAM that is non-volatile.

4-42 ________________ is a non-volatile memory type that exhibits the characteristics of read/write memory.

4-43 List some disadvantages of core memory.

4-44 PC memories can be visualized as an array of single unit cells, each of which can store information in the form of ________ or ________.

4-45 Each cell in the memory can store one ________ of information.

4-46 The ON/OFF information stored in a memory cell is known as ________________.

4-47 True/False. Memory capacity can be expanded in all PCs.

4-48 The abbreviation for memory, in terms of (K), represents ________________ locations whether bits, bytes, or words.

4-49 A PC system has an 8-bit microprocessor with a 16-bit memory structure. How many bits will the memory have if it has a memory capacity of 4K?

a- 4,096 bits
b- 32,768 bits
c- 65,536 bits
d- 131,072 bits

4-50 Select the decimal and octal representation for the maximum word address of a memory system that has 2K words or word number of:

Decimal	Octal
a- 2048	a- 1777
b- 2047	b- 2777
c- 2000	c- 3777
d- 2049	d- 4777

4-51 ______________ refers to the amount of memory locations required to store each PC instruction.

4-52 Determine the memory requirements (in units of K) for a program that has 10 rungs with 8 contacts per output and 20 rungs with 6 contacts per output. The memory requirement to store each contact is three bytes while it takes two bytes to store a coil output. The system memory has 8-bit words. 30 percent memory for future expansion after the minimum memory requirement is determined in 1/8K intervals.

4-53 Typical word lengths used in PCs are:

a- 8 bits
b- 12 bits
c- 16 bits
d- Varies with the controller

4-54 True/False. Data manipulation and other instruction require more memory to store the instruction.

4-55 Memory organization defines how certain areas of memory are used and is formally referred to as a ______________.

4-56 The executive and scratch pad are ______________ to the user when programming.

a- Important
b- Programmable
c- Transparent
d- Accessible

4-57 The ________________________ memory stores programmed instructions and data that is utilized by the processor to perform its control functions.

4-58 All data is stored in the ________________________ while programmed instructions are stored in the area allocated for the ________________________ program.

4-59 True/False. The input table is an array of bits that stores the status of digital inputs connected to the input interfaces.

4-60 True/False. Each connected input has a bit in the input table that corresponds exactly to the terminal where the input is connected.

4-61 True/False. The input table does not change the status of its bits while running a program.

4-62 A controller with a maximum of 128 outputs would require an output table of:

a- 128 bits
b- 128 bytes
c- 64 bits
d- Depends on how many are connected

4-63 A controller with 64 I/O and capacity for 128 I/O would require a combined I/O table of:

a- 128 bytes
b- 64 bits
c- 16 bytes
d- 64 bytes

4-64 The status of the bits in the output table is controlled by the ________________________ as it interprets the control program.

4-65 The output table is updated according to the control logic of the program and is updated during the ________________________ scan.

4-66 List some names that are used interchangeably with internal storage bits.

4-67 What is the purpose of using internal outputs?

4-68 Sketch a memory map indicating the beginning and ending word addresses of each section for a PC with the following specifications.

a- The total application memory is 2K.
b- The maximum amount of inputs is 128.
c- The maximum amount of outputs is 128.
d- The word length of the memory is 16 bits.
e- There are 64 internal outputs.
f- The system has 64 data registers.
g- The system uses the octal numbering system.
h- The user memory (for the control program) uses the rest of the 2K available.
i- The order in the memory is as follows:

1- Input table
2- Output table
3- Internal storage
4- Register storage
5- User program area

4-69 How much memory (in units of K) is left for the user program in problem 4-68?

4-70 In what type of format can the data in registers be stored?

4-71 True/False. The value equivalent to an analog input signal can be stored in a holding register.

4-72 Specify which of the following are constants (C) and which are variables (V).

____ Timer preset
____ Analog input
____ BCD output
____ Setpoints
____ Counter accumulated value
____ ASCII messages

4-73 True/False. The addresses of inputs and outputs are also specified in the user program memory.

4-74 The interpretation of the user program is accomplished by the processor's execution of the ______________________ program.

4-75 True/False. The power supply does not play an important role in the PC system operation.

4-76 True/False. PCs only accept voltages of 110 VAC or 220 VAC.

4-77 An important specification of the power supply is:

a- Percentage variation in line conditions
b- Current supply
c- Operating voltage
d- All of the above

4-78 What happens when the power supply detects an over- or under-voltage condition?

4-79 The rating (in VA) of a constant voltage transformer should be selected based on:

a- Average power loading
b- Worst-case power loading
c- Primary rating
d- Best-case power loading

4-80 True/False. Constant voltage transformers that do not filter harmonic are recommended for PC applications.

4-81 When are isolation transformers needed?

4-82 True/False. Power supply overloading is a problem that can be easily detected.

4-83 What should be done if the summation of current requirements for an I/O configuration is greater than the total current supplied by the power supply?

4-84 True/False. The executive program and supporting software used by the processor are also called the system memory.

UNIT 5

THE INPUT/OUTPUT SYSTEM

- The I/O system provides the physical connection between the field devices and the central processing unit.

DISCRETE INPUTS/OUTPUTS

- The most common type of I/O interface is the discrete type.
  Standard ratings for discrete I/O are as follows:

Inputs	Outputs
24 Volts AC/DC	12-48 Volts AC
48 Volts AC/DC	120 Volts AC
120 Volts AC/DC	230 Volts AC
230 Volts AC/DC	TTL level
TTL level	120 Volts DC
Non-voltage	230 Volts DC
Isolated inputs	TTL level
	Isolated outputs
	Contact (relay)

- A logic 1 indicates ON or CLOSED; a logic 0 indicates OFF or OPEN.
- The input circuit of an AC/DC input module generally consists of a power section and a logic section separated by an isolation circuit.
- The input module (through the bridge circuit) converts the input signal to a low-level DC signal and is then passed through a filter delay.
- Electrical isolation is provided so that there is no electrical connection between the power and logic sections.
- TTL (5 volts DC) I/O allows interfacing with TTL compatible devices including solid-state controls and sensing instruments. These interfaces also have an input delay circuit; however, the delay is much shorter than the AC/DC interfaces.
- Nonvoltage interfaces allow sensing of the closure of dry contacts.
- AC outputs are composed of the logic and power sections separated by an isolation circuit.
- The switching section of an AC output generally uses a triac or silicon controlled rectifier (SCR) to switch the power to the field device. Protection at the output is accomplished by the use of an RC snubber and a metal oxide varistor (MOV).
- DC output interfaces generally use a power transistor to switch the DC voltage. Protection at the outputs is achieved by the use of a free-wheeling diode.

- The contact output modules are used to switch AC or DC signals to the loads. These interfaces have applications in the multiplexing of analog signals to drives, switching small currents at low voltages, and switching high currents (high power contacts).

- Isolated I/O interfaces have a common return line for each group of inputs or outputs. These modules are used when it is required to have input or output field devices in different common return lines or power sources.

NUMERICAL DATA INPUT/OUTPUT INTERFACES

- From the hardware point of view numerical data I/O interfaces came about to support the new capabilities of arithmetic operation and data manipulation achieved by the new microprocessor technology.

- Numerical I/O is classified in multibit and analog interfaces. Multibit interfaces allow a group of bits to be input or output to devices that handle several bits at a time in parallel (BCD) or in serial form (pulse inputs or outputs).

- Analog I/O allows monitoring of continuous analog voltages or currents compatible with sensors, motor drives, and process instruments.

- Analog interfaces are available in unipolar (positive) and bipolar (positive and negative) ratings. Standard ratings found in analog I/O modules are as follows:

Analog Inputs	Analog Outputs
4-20 mA	4-20 mA
0 to +1 Volts DC	10-50 mA
0 to +5 Volts DC	0 to +5 Volts DC
0 to +10 Volts DC	0 to +10 Volts DC
1 to +5 Volts DC	2.5 Volts DC
5 Volts DC	5 Volts DC
±10 Volts DC	±10 Volts DC

- Analog inputs use an analog-to-digital converter (ADC) to transform the signal to an equivalent digital value proportional to the analog signal.

- Analog input modules have a high input impedance that allows interfacing with high source-resistant outputs from input devices.

- Analog outputs use a digital-to-analog converter (DAC) to transform a numerical value to an equivalent analog value.

- Register I/O modules provide parallel communication between an input or output device and the processor. Typical field devices that use these modules are thumbwheel switches and seven-segment LED indicators.

- The encoder/counter input provides a high-speed counter external to the processor. Signals used in the interface include marker, limit switch, channel A, channel B, ACC = PRE, and ACC > PRE.

SPECIAL I/O INTERFACES

- Special I/O modules may incorporate an on-board microprocessor that can perform processing tasks independent of the CPU (distributed processing).

- Fast input modules detect very fast pulses and maintain the status for one scan (pulse stretcher). Typical devices that may be used with this interface include proximity switches, photoelectric cells, and instrumentation equipment.

- The ASCII interface is used for sending and receiving alphanumeric data between the controller and peripheral equipment. Some ASCII modules may include an on-board microprocessor that will speed up the transmission rate.

- Stepper motor interfaces provide output pulses that are compatible with stepper motor translators. The pulses represent distance, rate, and direction commands to the motor. Acceleration and deceleration are determined by the rate of output pulses.

- Servo interfaces are used in applications to perform motion control. Advantages of servo control include shorter positioning time, higher accuracy, better reliability, and improved repeatability in the coordination of axis motion.

- PID modules provide proportional-integral-derivative action in applications requiring closed-loop control. PID is often referred to as a three-mode closed-loop feedback control.

- Data processing modules are microprocessor-based interfaces that allow data handling functions typically performed by the CPU or a small dedicated computer. These functions include storing and retrieving recipes, storing and displaying operator messages, generating production reports, and other processing jobs. These interfaces relieve the main CPU from the data handling burden and help eliminate a large amount of data processing programming in the main CPU.

- Network interface modules allow communication among several PCs over a high-speed local area network.

REMOTE I/O

- Remote input and output systems are used in applications where the total system is spread out in a plant location. The use of remotes not only simplifies control but also reduces tremendously the amount of wiring.

- Remote racks are connected to the main CPU in a daisy chain or star configuration.

REVIEW QUESTIONS

5-1 The I/O system provides the interface between:

a- Field equipment and output modules
b- Field equipment and CPU
c- Input modules and CPU
d- Input modules and output modules

5-2 What is the most commonly used type of I/O interface?

a- Analog
b- Discrete inputs
c- Output modules
d- Discrete I/O

5-3 Characteristics of discrete I/O interfaces limit their use to field devices that are:

a- ON/OFF
b- OPEN/CLOSE
c- Switch closures
d- All of the above

5-4 Name five discrete input devices and five discrete output devices that are interfaced with discrete I/O.

5-5 List five standard ratings for both discrete inputs and outputs.

5-6 A logic 1 indicates ________________, and a logic 0 indicates ________________.

5-7 The input circuit of an AC/DC input module is composed of a ________________ and a ________________ section.

5-8 AC/DC input modules may also include:

a- Fuse
b- AC detection circuit
c- Isolation circuit
d- Bridge rectifier

5-9 The bridge rectifier section in the input module converts the input signal to:

a- A DC level
b- A low AC level
c- A reduced noise level signal
d- An isolated DC signal level

5-10 Typical input signal delays range from ______________ to ______________ msec.

5-11 The input signal is recognized as a valid input if the signal:

a- Exceeds the threshold voltage
b- Stays ON for one scan
c- Exceeds the threshold voltage and remains there for at least the filter delay
d- Stays ON within 10% of the voltage threshold

5-12 Electrical isolation is provided so that there is no electrical connection between field devices and the controller and is generally provided by ______________ or ______________.

5-13 Reference Fig. 5-1. If a limit switch input is NO, the power indicator in the input module should be ______________.

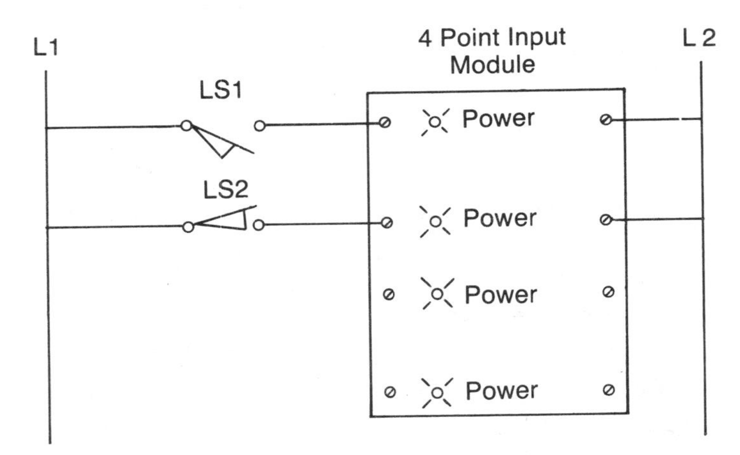

Figure 5-1. Problem 5-13.

5-14 Reference Fig. 5-1. If a limit switch input is NC, the power indicator in the input module should be ______________.

5-15 TTL input modules receive TTL signals from devices such as ______________ (name one).

5-16 Is the input delay caused by TTL inputs generally longer or shorter than the AC/DC inputs?

5-17 Non-voltage input modules are used to recognize:

a- 110 volt signals
b- TTL signals
c- Closures of dry contact inputs
d- Low-level DC signals

5-18 Sketch the three sections that make up an AC output module (indicate signals).

5-19 The switching section of an AC output module generally uses a ________________ to switch the power.

5-20 The two most common protection elements of an AC output module are ________________ and ________________.

5-21 The switching elements in DC output modules are ________________ and generally protected by ________________.

5-22 Name two applications in which contact output modules are used.

5-23 Name one output device that is generally driven by TTL output modules.

5-24 Why are isolated I/O modules sometimes used?

5-25 In Fig. 5-2, properly connect the input/output devices to the modules shown.

5-26 Numerical data I/O interfaces are a product of new capabilities in ________________ and ________________ instructions.

5-27 Numerical data I/O interfaces can be categorized in two major groups: ________________ and ________________.

5-28 Indicate whether the following numerical I/O devices are an analog input, analog output, multibit input, or multibit output. (i.e., AI, AO, MI, MO)

____ Pressure transducer
____ Chart recorders
____ Encoders
____ Potentiometer
____ Thumbwheel switches
____ Seven-segment display

5-29 Analog interfaces can be found in ________________ or ________________ type ratings.

5-30 List at least five standard analog input ratings and five analog output ratings.

5-31 Analog input interfaces use an analog-to-digital converter (ADC) to convert the analog input signal into a value that is ________________ to the input signal.

5-32 The input impedance of an analog interface in general is ________________, which allows them to interface to ________________ outputs from input devices.

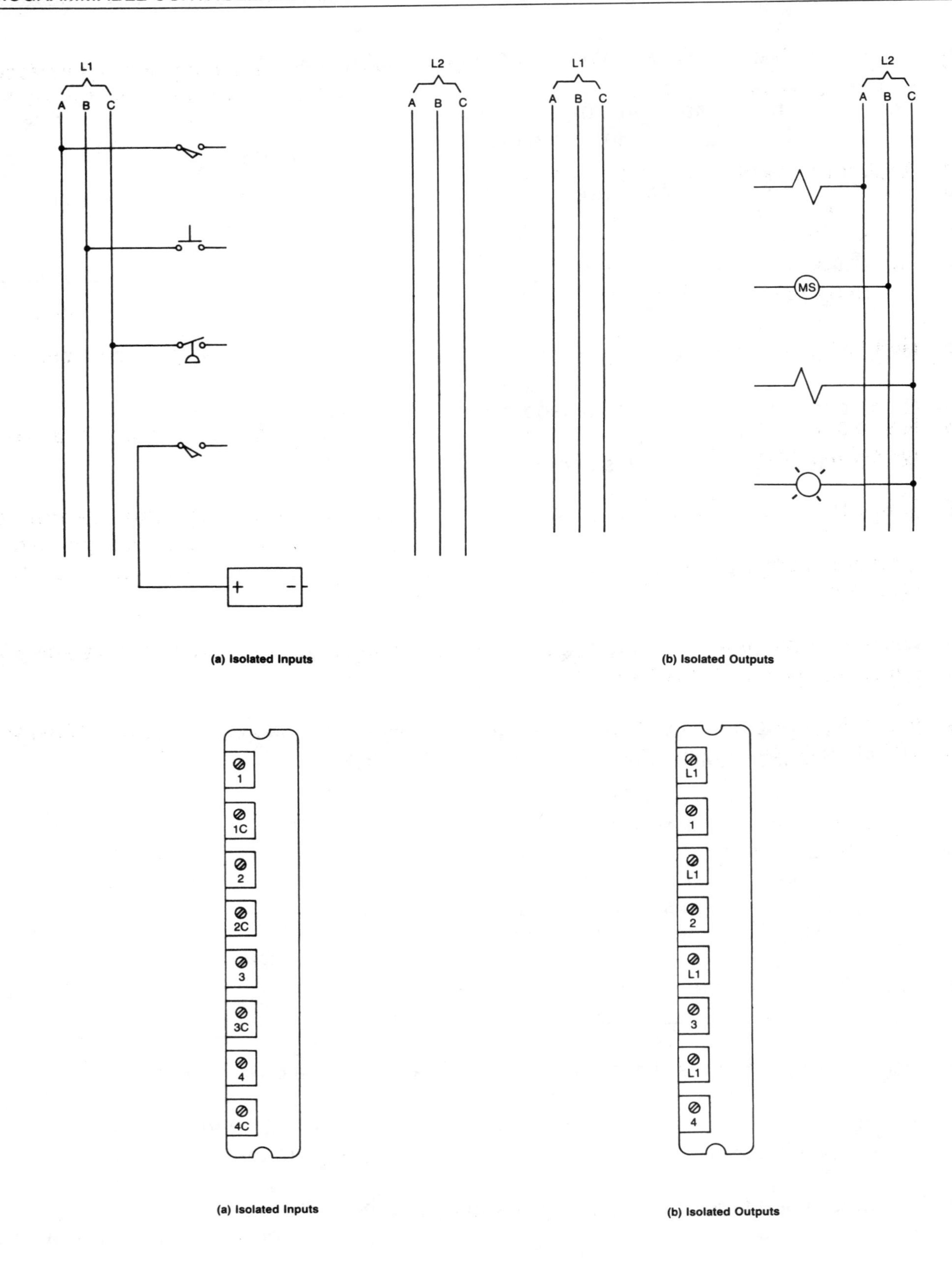

Figure 5-2. Problem 5-25.

5-33 ________________________ cables are generally used when connecting analog interfaces to field devices, and their ground wires are connected at one end located at the ________________________.

5-34 Analog values after conversion can range from:

a- +1 to +10 volts
b- -10 to +10 volts
c- 0000 to 9999 BCD or 0 to 32767 decimal
d- 0 to 32767 in decimal

5-35 Register input modules provide ________________________ communication between ________________________ and ________________________.

5-36 Name one example of an input device interfaced with a register input module and give a typical application for such an interface.

5-37 Register input modules generally accept voltages in the range of ________________________ to ________________________ volts DC and are grouped in modules containing ________________________ or ________________________ input channels.

5-38 Name one example of an output device interfaced with a register output module and give typical application for such an interface.

5-39 In Fig.5-3, identify the least and most significant bits of the register, the 1s, 10s, 100s, and 1000s units, and the most significant bit of the 100s unit.

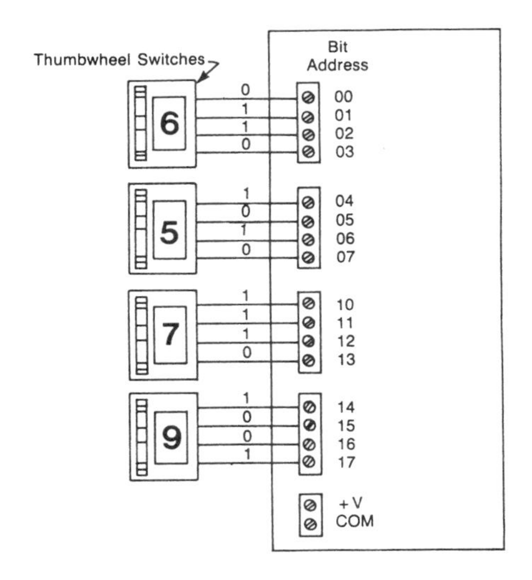

Figure 5-3. Problem 5-39.

5-40 An encoder input module can also be considered a:

a- Special I/O module
b- High-speed counter
c- High-speed counter external to the processor
d- Interface to an operative console

5-41 Name two cases in which the encoder input module can be applied.

5-42 Explain briefly the operation of an incremental encoder when connected to an encoder input module. How does this differ from an absolute encoder?

5-43 Which of the following is not a typical input connection to an encoder input module?

a- Channel A
b- Channel B
c- Power
d- Marker
e- Reverse count
f- Limit switch

5-44 Special I/O modules are also known as ______________________ modules.

5-45 Some special I/O modules have added intelligence by utilizing an on-board ______________________.

5-46 Special I/O modules are generally available in ______________________ to ______________________ sized programmable controllers.

5-47 Name an example of a thermocouple input interface and its function.

5-48 In Fig. 5-4, properly connect the thermocouple input to the module (specify the cable and ground connections).

5-49 The operation of a fast input module can be thought of as a ______________________ that allows a signal to remain valid for one scan.

5-50 Which of the following signals cannot be interfaced using a fast input module?

a- 12 volts DC
b- 110 VAC
c- 24 volts DC
d- 10 volts DC

5-51 Name two devices that are generally interfaced to the PC with a fast input module.

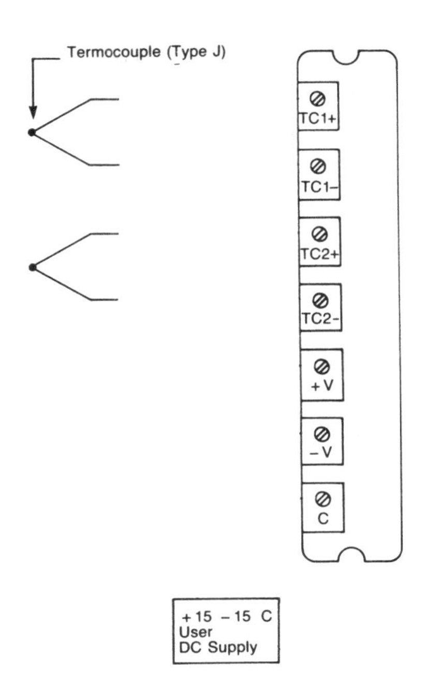

Figure 5-4. Problem 5-48.

5-52 Describe the application of the ASCII module and typical devices it is used in conjunction with.

5-53 Which of the following is not a standard communication link for an ASCII serial communication module?

a- RS-232C
b- RS-422
c- Parallel
d- 20 mA loop

5-54 If the ASCII module contains an on-board microprocessor, the communication is ______________ than directly from the main CPU.

5-55 In a non-smart ASCII module, the characters are not received and sent:

a- One byte at a time
b- Character by character
c- Every two characters
d- On interrupt basis

5-56 Assume that a PC has a total scan time of 3.5 msec per K of program, and the control program takes 4K of memory plus 1K of data storage; the communication speed at the device can be selected for 300, 600, 1200, 1500, or 2400 baud. What is the maximum speed the PC would be able to communicate? Show your work.

5-57 From the previous problem, if the I/O update has an extra 2 msec overhead for each scan, what would be the maximum communication speed?

5-58 Name at least three parameters that must be taken into consideration to ensure proper transmission when using the ASCII communications module.

5-59 The strain gage module is used to interface devices that: (choose the best answer)

a- Employ strain gage circuits
b- Have output signals of very low level
c- Have strain gage circuits that provide low-level output signals
d- Do not have output amplifiers

5-60 The strain gage module can be used to interface:

a- Pressure transducers
b- Load cells
c- Pressure switches
d- a and b
e- a and c

5-61 The stepper motor interface is an ______________________ module that is used to generate ______________________ compatible with stepper motor translators.

5-62 The pulses sent to a stepper translator from the stepper motor interface generally represent:

a- Distance
b- Rate
c- Direction
d- All of the above

5-63 The acceleration or deceleration of the stepper motor is determined by the ______________________ of output pulses.

5-64 Servo interfaces are used in applications requiring:

a- Point-to-point control
b- Clutch-gear systems
c- Axis positioning
d- a and c
e- a and b

5-65 List at least three advantages that servo control has over clutch-gear systems in performing motion control.

5-66 In brief terms, describe the operation of the servo interface module.

5-67 The axis positioning interface is another special module that is used for ______________.

5-68 PID control is also referred to as:

a- Proportional-integral-differential control
b- Proportional-integral-derivative control
c- Three-mode closed-loop control
d- Set-point control

5-69 What is the basic function of a PID control module?

5-70 Name at least three process variables that are typical inputs to a PID module.

5-71 In the following equation, identify the term E, Kp, Ki, and Kd and explain their main functions.

$$Vout = K_P E + K_I \int E\,dt + K_D \frac{dE}{dt}$$

5-72 Match the following terms with the appropriate answer.

a- Update time	____ Quantity compared to the error signal
b- Error dead band	____ Desired output
c- Set point	____ Rate or period of update
d- Square root extraction	____ Linearized scaled output
e- Derivative gain	____ Reset action
f- Integral gain	____ Rate time
g- Kd = KpTd; Td is:	
h- Ki = Kp/Ti; Ti is:	

5-73 In the following block diagram, fill in the blanks with the appropriate terms.

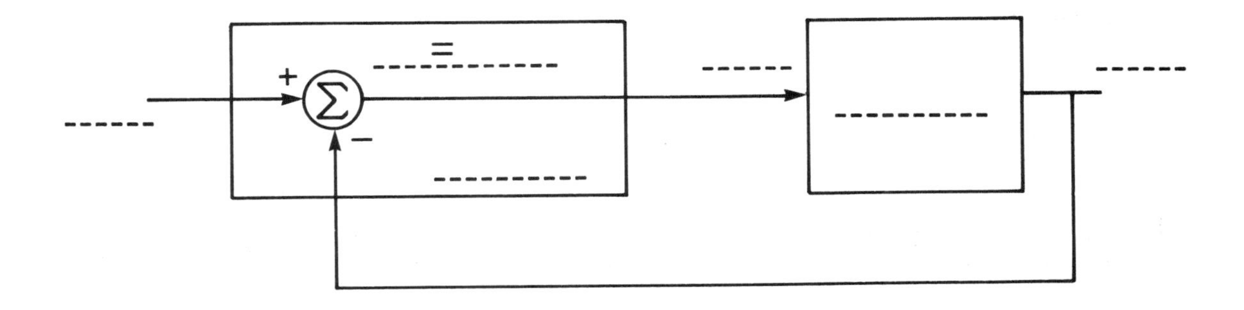

5-74 Explain the main functions performed by data processing modules and how these modules can be used.

5-75 Explain the function and operation of network interface modules.

5-76 What are the typical configurations for I/O subsystems in remote configurations? Sketch these configurations.

5-77 Name some advantages of using remote I/O systems.

5-78 Match each of the following specifications with the appropriate answer:

a- Input voltage rating	____ Duration at which input signal must be ON after threshold voltage is passed
b- Input threshold voltage	____ Maximum leakage of output modules when they are OFF
c- Input delay	____ Response time for output to go from ON to OFF
d- Output current rating	____ Number of I/O circuits in a module
e- Surge current	____ Maximum current that an output circuit can carry under load
f- Off-state leakage current	____ Nominal AC or DC voltage that specifies magnitude of signal accepted
g- Output off-delay	____ Voltage isolation between logic and power circuits
h- Digital resolution	____ Voltage level at which input signal is recognized as being ON
i- Points per module	____ Maximum current duration an output module can withstand
j- Ambient temperature rating	____ Definition of how close the converted analog signal approximates
k- Electrical isolation	____ Maximum operating temperature

UNIT

--]6[--

PROGRAMMING AND PERIPHERAL DEVICES

PROGRAMMING DEVICES

- The most common peripheral devices available for PC programming include:
  - Cathode ray tube (CRTs)
  - Miniprogrammers
  - Program loaders
  - Memory burners
  - Computers
- CRTs are the most common devices used for programming the programmable controller. CRTs offer the advantage of displaying large amounts of logic on the screen.
- CRTs are classified in two categories: dumb and intelligent. Dumb CRTs communicate with the PC for on-line programming, while the microprocessor-based CRTs (intelligent) can be used on-line as well as off-line programming.
- Miniprogrammers, also known as hand-held or manual programmers, are an inexpensive and portable means for programming small PCs.
- Miniprogrammers are a useful tool for start-up, changing, and monitoring the control logic.
- Program loaders are used for loading and reloading the control program into the PC memory.
- Typical program loaders include cassette recorders and electronic memory modules.
- Memory burners are used to enter a program into a memory chip by coding the program and then blasting onto the memory chip to store the program.

OTHER PERIPHERALS

- Several peripheral devices have been developed to improve communication between the PC system and the operators, and they fall into four major categories: data entry, documentation and reporting, displays, and control devices.
- Data entry devices, such as thumbwheel switches and operator interface consoles, are the most common devices used as a means of entering data into the controller. These devices are interfaced in general with register input modules.
- Line printers and report generation systems are used to produce reports of production and system operation. Line printers are interfaced with ASCII communication modules.
- The most commonly used display devices include seven-segment, alphanumeric, and color-graphic displays. These display units allow the operator to view data directly in the device that may indicate process parameters, alarm indications, and a total view of the whole process.

- Control devices are a type of control-station equipment that is used between the PC system and the final field- controlled element. These stations are useful for back-up of digital as well as analog signals and for system start-ups.

PERIPHERAL INTERFACING

- Communication standards fall into two major categories: proclaimed and de facto. Proclaimed standards have been officially established while de facto standards have been adopted and gained popularity through use without official definition.

- Some proclaimed standards include IEEE 488 bus, EIA RS-232C, and EIA RS-422. Example of de facto standards are the PDP-11 Unibus, and 20mA current loop.

- The RS-422 standard is a balanced link communication that can achieve greater communication distances at faster baud rates than the RS-232C (unbalanced link).

REVIEW QUESTIONS

6-1 Name five of the most common peripheral devices used with programmable controllers.

6-2 True/False. CRTs are the most commonly used PC programming devices.

6-3 What is the greatest advantage CRTs have over small or miniprogrammers?

6-4 CRTs can be classified as:

a- Dumb
b- Intelligent
c- Smart
d- All of the above

6-5 Non-smart CRTs are used for programming the controller:

a- On-line
b- Off-line

6-6 Characteristics of smart CRTs are :

a- Microprocessor-based
b- Off-line programming
c- On-line programming
d- All of the above

6-7 Describe on-line and off-line programming.

6-8 List three uses of mini-programmers.

6-9 True/False. All miniprogrammers are not microprocessor- based.

6-10 Describe the function of program loaders and name two.

6-11 Memory burners are used for programming ROM, PROM, and EPROMS for permanent program storage. The programs are generally coded and debugged utilizing a read/write-based programming device and then ______________________ onto memory chips.

6-12 Name four categories of peripheral devices that are used for communications and control purposes in programmable controller systems.

6-13 Name typical applications in which thumbwheel switches (TWS) are used.

6-14 How many wires are connected to the PC register module in a three-digit thumbwheel switch?

a- 4 wires
b- 12 wires
c- 10 wires
d- 3 wires per digit

6-15 Thumbwheel switches provide information to the PCs in ______________________ format.

6-16 A numerical data entry panel can be considered:

a- A multi-function data entry device
b- An operator's debugging tool
c- An operator's control panel entry device
d- a and b
e- b and c
f- a and c

6-17 Name at least two cases in which line printers are utilized as part of the PC control system.

6-18 What type of output module is generally required with seven-segment indicators? How many wires are required per digit?

6-19 Explain intelligent alphanumeric displays and how they can be used.

6-20 Which of the following is not true for colorgraphic displays?

a- Used for alarm displaying
b- Used for peripheral documentation
c- Used for message displays
d- Used for process description in graphic form

6-21 What is the main purpose of manual control stations, and when are they useful?

6-22 Identify the following standards as either (P)- proclaimed or (D)- de facto standards:

____ EIA RS-232C
____ Unibus
____ IEEE 488
____ 20 mA current Loop
____ EIA RS 422

6-23 ______________________ information is generally sent to/from the PC peripheral equipment in a serial format.

6-24 Data communications links can be:

a- Unidirectional
b- Serial RS-232C
c- 20 mA current loop
d- All of the above

6-25 Data sent in either direction, but only in one direction at a time, is known as ______________________.

6-26 Data sent in both directions simultaneously through two lines is known as ______________________.

6-27 The RS-232C standard connector has ______________________ possible signal lines.

6-28 Describe three electrical characteristics of the RS-232C standard.

6-29 The ______________________ communication standard can achieve greater distances between communicating devices.

a- RS-232C
b- 20 mA current loop

6-30 Which of the following is not true in an ASCII pulse train (one character)?

a- Has a start and stop bit
b- ASCII information is in seven bits
c- Has a total of 10 bits
d- There can be checksum bit for error detection

6-31 Match the following:

____	Unbalanced link	a- Mechanical specifications for RS 422
____	RS-232C	b- RS-232C
____	Balanced link	c- Maximum data rate 19.6K baud
____	RS-449	d- RS-422
____	Logic 1	e- 110 baud
____	Logic 0	f- Mark
____	1 stop bit	g- Space
____	2 stop bits	h- 300 baud

6-32 Sketch the relationship between distance and data rate for the RS-232C and RS-422 as illustrated in Fig. 6-1 (with and without termination).

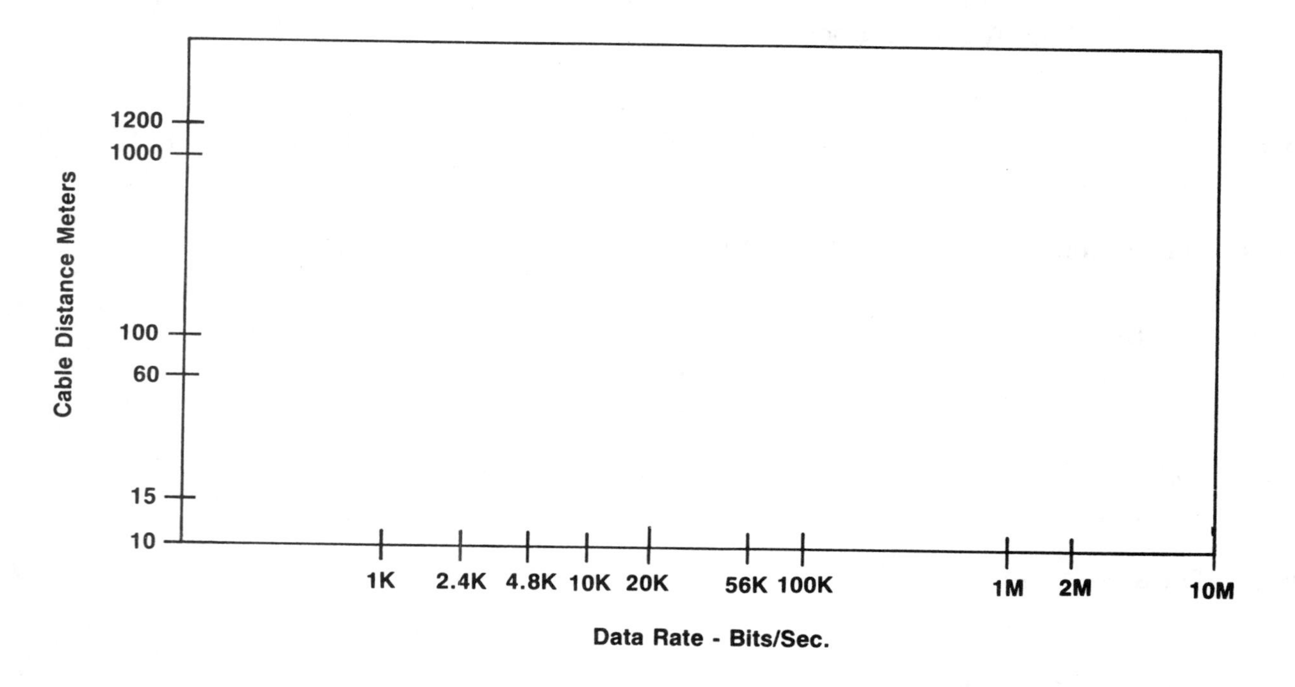

Figure 6-1. Problem 6-32.

6-33 Which of the following wires is not present in a 20 mA current loop?

a- Transmit plus
b- Receive minus
c- Receive plus
d- Signal ground

UNIT

--]7[--

PROGRAMMING LANGUAGES

- The four most common languages encountered in programmable controllers are:

  - Ladder diagrams
  - Boolean mnemonics
  - Functional blocks
  - English statements

- Basic PC languages include ladder diagrams and Boolean mnemonics, while functional blocks and English statements are considered high level languages.

- Functional blocks can be considered an extension of ladder diagrams, but the software capabilities associated with functional blocks are much greater.

- PC instructions can be categorized in the following six operations:

  - Relay logic
  - Timing and counting
  - Arithmetic
  - Data manipulation
  - Data transfer
  - Flow of control

- Relay logic symbology usually is represented by contacts (NO and NC) and coils. These symbols are used to generate what is called a logic or ladder rung.

- A rung is said to have continuity if there is a continuous path from left to right (logic continuity) in the contact elements that drive an output coil.

- A normally open contact instruction examines for an ON or CLOSED condition while the normally closed contact examines for an OFF or OPEN condition in order to provide logic continuity in a ladder rung.

- Software timers and counters provide the same function as hardware timers and counters. However, these software instructions are more accurate and less likely to fail because they are incorporated in the controller.

- The basic arithmetic operations implemented in PCs are addition, subtraction, multiplication, and division. These instructions operate on registers or words located in the memory system of the PC.

- Data manipulation instructions are an enhancement of the basic ladder diagram instruction set. These instructions include comparisons of registers, logic matrix operations, and data conversion.

- As the name implies, data transfer instructions deal with the transfer of information between registers or words. These instructions can be performed on a register to register basis or in a group of registers called a block transfer.

- Sequencers or drum instructions are used to replace hardware sequences or drum cylinder type mechanisms.

- Boolean mnemonics use the Boolean algebra theory to develop logic statements used in control sequence. Boolean algebra was developed in the mid 1800s by George Boole.

- English statement languages are types of control statements that perform functions similar to those of functional blocks. However, this language follows the programming structure of a BASIC-type language used in computers.

REVIEW QUESTIONS

7-1 ______________________ is a symbolic instruction set that is used to create a PC program.

7-2 True/False. The ladder diagram instruction set is often referred to as contact symbology.

7-3 What is a ladder rung?

7-4 ______________________ and ______________________ are the basic symbols of the ladder diagram instruction set.

7-5 Name the six most common instruction types available in PCs.

7-6 In ladder diagram programs, all outputs are represented by:

a- Contact symbols
b- Coil symbols
c- a or b
d- None of the above

7-7 List four programming languages used in PCs.

7-8 Which of the following may be considered high-level languages: (circle all that apply)

a- Functional blocks
b- Boolean mnemonics
c- Ladder diagrams
d- English statements

7-9 What is the difference between basic PC instructions and high-level PC instructions?

7-10 True/False. A PC can utilize more than one language.

7-11 What is logic or rung continuity?

7-12 True/False. The ladder diagram language allows only relay-type instructions.

7-13 For what purpose is the energize coil instruction used?

7-14 Describe how a latched coil condition is reset or unlatched.

7-15 What must occur to a referenced contact address to retrigger a transitional contact or one-shot instruction?

7-16 Describe the differences between the TON and TOF instructions.

7-17 The up-counter increments its accumulated value each time the up-count event makes an:

a- ON-to-OFF transition
b- OFF-to-ON transition
c- a or b

7-18 If a given PC does not have the TOF instruction, how would the TON instruction be implemented to do the same function?

7-19 What is the function of the retentive timer reset (RTR)?

7-20 Describe the operation of the program in Fig. 7-1.

100 R 145 MUL R 130 = R 135 50

Figure 7-1. Problem 7-20.

7-21 Describe the MCR instruction and give examples of how it is used.

7-22 Describe the operation of the program in Fig.7-2.

7-23 Which of the following sequences of closed contacts would not energize the output coil in Fig.7-3. Redraw the circuit to accommodate all possible sequences.

a- Contacts 5,6,and 7 CLOSED
b- Contacts 5,6,8,and 9 CLOSED
c- Contacts 10,11,8,and 7 CLOSED
d- Contacts 10,11,and 9 CLOSED

7-24 From the timing diagram in Fig.7-4, determine which are the TON delay energize and TON delay de-energize instructions?

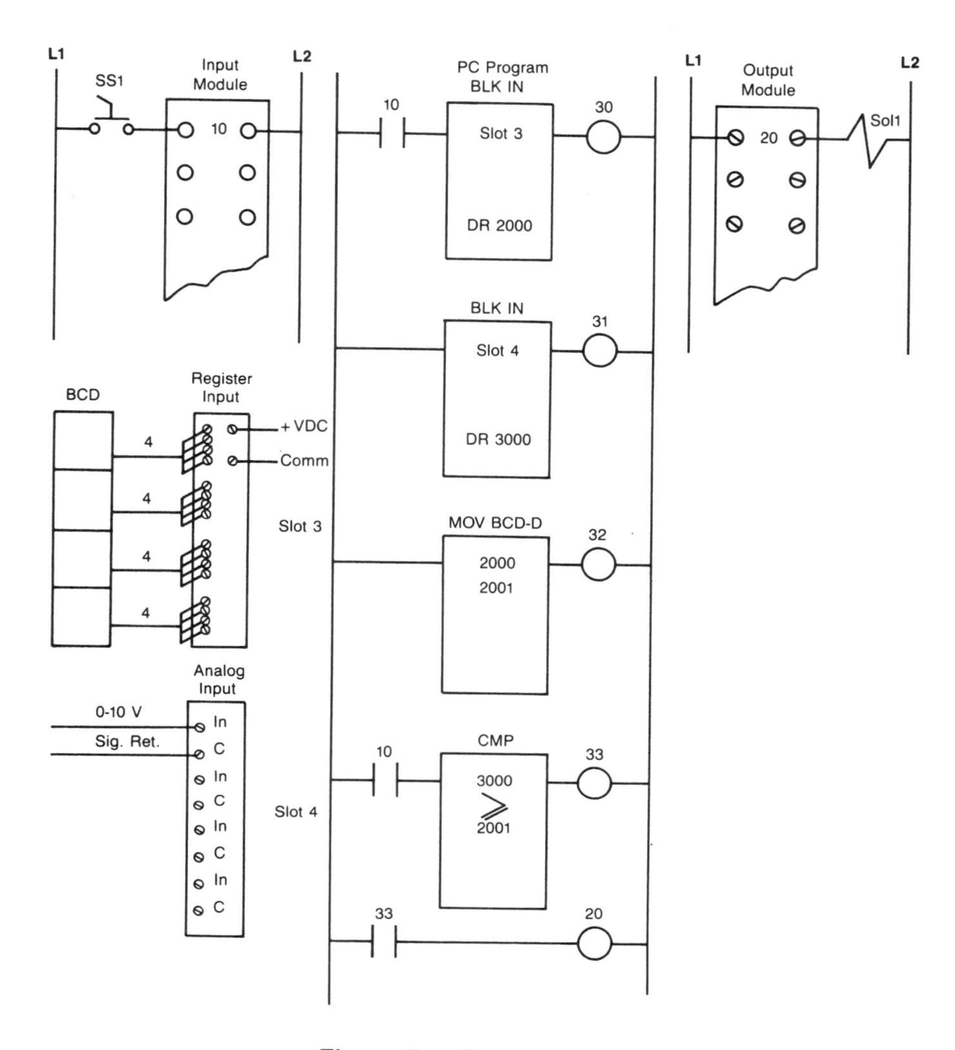

Figure 7-2. Problem 7-22.

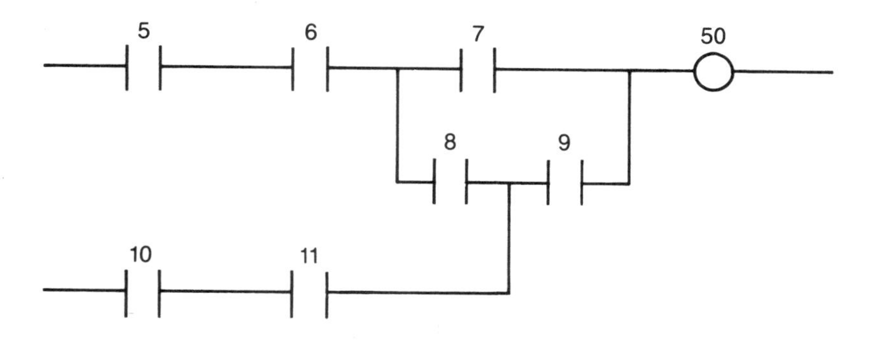

Figure 7-3. Problem 7-23.

a–

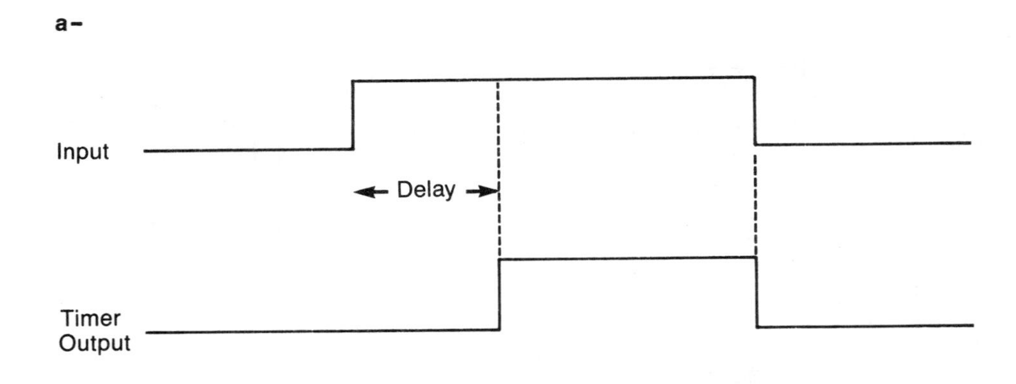

b–

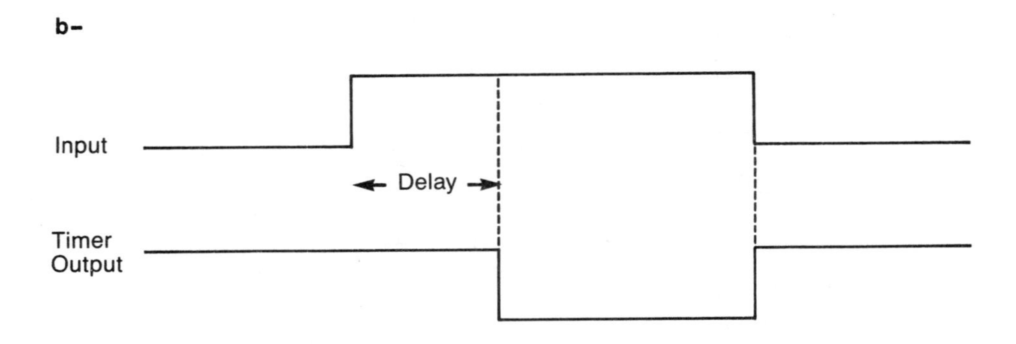

Figure 7-4. Problem 7-24.

7-25 Write the sequences using Boolean instructions for each of the ladder programs in Fig. 7-5.

a–

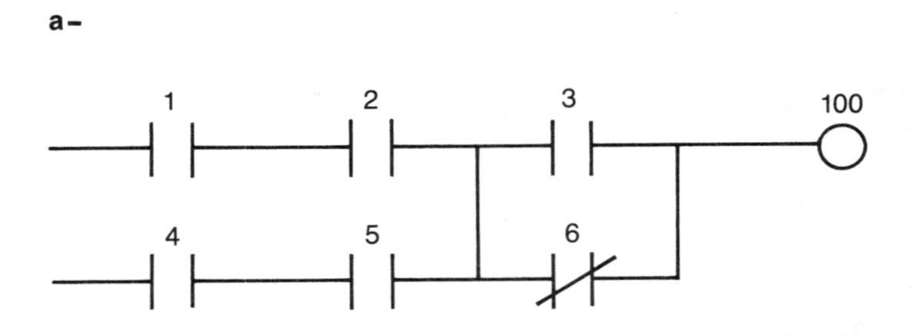

Figure 7-5. Problem 7-25.

b-

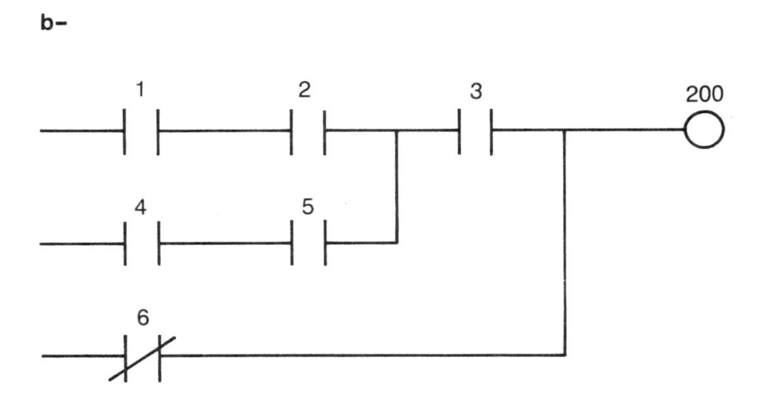

c-

1 2 3 4 250

7

5 6

Figure 7-5 continued. Problem 7-25.

7-26 Write the ladder diagram programs that correspond to the following Boolean programs.

a-
LD 1
OR 3
LD 2
OR 4
AND LD
LD 5
OR 7
AND 6
OR LD
OUT 300

b- LD 1
AND 2
LD 4
AND 5
OR LD
OR 7
LD 3
OR NOT 6
AND LD
OUT 350

c- LD NOT 1
AND 2
LD 4
OR 7
AND 5
OR LD
LD 3
OR NOT 6
AND LD
OR NOT 8
OUT 450
AND 9
OUT 500
AND NOT 10
OUT 600

7-27 ______________________ are timers that will stop timing when logic continuity is lost and will resume timing starting at the ______________________ when the control line is energized again.

7-28 Counters are used to count events or occurences of inputs when a transition from OFF-to-ON occurs, which implies that they are ___________ triggered.

7-29 Is there any advantage to having an addition instruction in functional block form instead of relay-type instruction as illustrated in Fig.7-6?

7-30 In some controllers, the subtraction instruction could also be used for the ______________________ instruction.

7-31 ______________________ math instructions are used when handling large numbers, and ___________ registers are used to store the result.

7-32 True/False. The compare functional block (not a matrix block) is used to compare the contents of two or more registers.

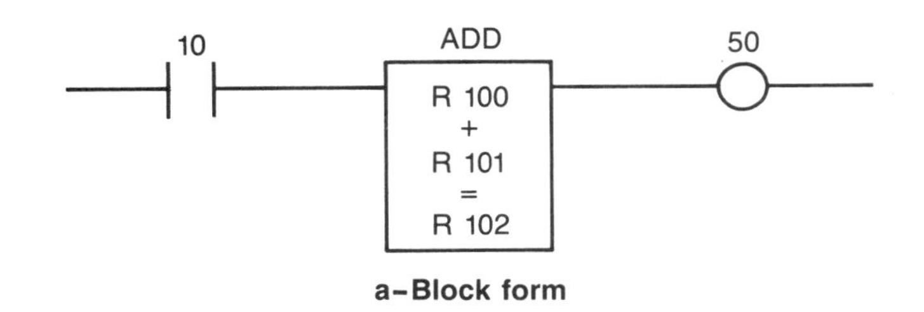

a-Block form

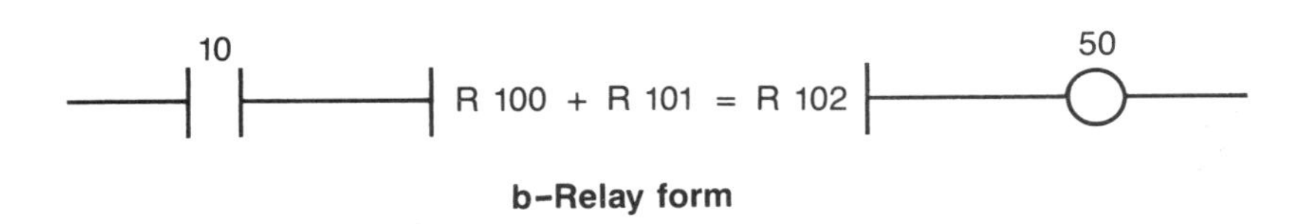

b-Relay form

Figure 7-6. Problem 7-29.

7-33 True/False. When the rotate right instruction is executed, the least significant bit of the register will be lost.

7-34 The ____________________ instruction checks for an ON or OFF status of a bit in a register or memory location.

a- Move block
b- Set constant parameters
c- Data comparisons
d- Examine bit

7-35 In which of the following instructions does the user specify certain parameters associated with the algorithm to control the process?

a- Sequences
b- Diagnostics
c- PID
d- None of the above

7-36 True/False. Users can purchase a PC and then select the right language for their application.

7-37 For users familiar with relay logic, the best PC language suited for a relay-replacing application would be:

a- Boolean
b- English statements
c- Ladder diagrams
d- Functional blocks

7-38 True/False. When selecting a PC language, scan time is a very important factor to be considered.

7-39 Name some typical conversion operations that are performed in data conversion block instructions.

7-40 Which of the following instruction types is used to perform AND, OR, NAND, EXCLUSIVE-OR, NOR, and NOT logic operations on two or more registers?

a- Data conversions
b- Data comparisons
c- Set constant parameters
d- None of the above

7-41 The ______________________ instruction causes a group of register or word locations to be copied from one location to another.

7-42 ______________________ instructions are useful in creating routines that signal the operator of a machine malfunction.

7-43 True/False. MCR instructions can be nested and overlapped in the same program.

7-44 Which of the following instructions is associated with data transmission of alphanumeric characters from a PC to a peripheral device?

a- FIFO stack transfers
b- Block transfer (IN/OUT)
c- ASCII transfers
d- Sequences

UNIT

--]8[--

PROGRAMMING THE CONTROLLER

- The first step in developing the Programmable Controller system is the definition of the control task, followed by the strategy definition and a systematic approach.

- The definition of the strategy includes the determination of the sequences of processing steps that must take place within a program to produce output control. This is also known as the development of an algorithm (control).

- A systematic approach follows a set of guidelines that are required to properly develop an organized system. Approach guidelines apply to two major types of projects: new applications and modernizations of existing equipment.

- Flowcharting is one technique used in planning a program after a written description is made.

- Address assignment of internals includes internal outputs, registers, timers and counters, and MCRs.

- Certain parts of the system are left hardwired for safety reasons. Elements such as emergency stops and master start push buttons should be left hard-wired so that a system can be disabled without PC intervention.

- Coding is the process of writing or rewriting the logic or relay diagram into PC ladder program form.

- The programming of a normally open or normally closed PC contact depends on the manner it is required to operate in the logic program.

- A normally closed input device is generally, but not always, programmed normally open if it is required to act as a normally closed contact.

- A programmed normally open PC contact tests for an ON or CLOSED condition at the input to close its contacts in the PC program.

- A programmed normally closed PC contact tests for an OFF or OPEN condition at the input to maintain its contacts closed in the PC program.

- Objectives of modernizing existing equipment include more reliable control system, less energy consumption, less space, and a flexible system that can be expanded.

REVIEW QUESTIONS

8-1 Which of the following steps should be the first taken towards achieving a properly designed PC control system?

a- Approach the system in a systematic manner
b- Flowchart the process
c- Define the control task
d- Define the strategy to be used

8-2 An ______________________ is a procedure that involves the determination of sequences or steps that take place in a program to produce the output control.

8-3 List four guidelines that are recommended as an approach to program design in a modernization project.

8-4 An existing ______________________ often defines the sequence of system operations to be controlled in a modernization project.

8-5 System operation for new applications usually begins with:

a-Sample diagrams
b-Specifications
c-Control strategy
d-Logic diagrams

8-6 ______________________ is one technique used in planning a control program.

8-7 Logic sequences for a control program can be created using:

a-Logic gates
b-Relay ladder symbology
c-PC contact symbology
d-All of the above

8-8 Draw the equivalent logic gate diagrams for the circuit in Fig. 8-1.

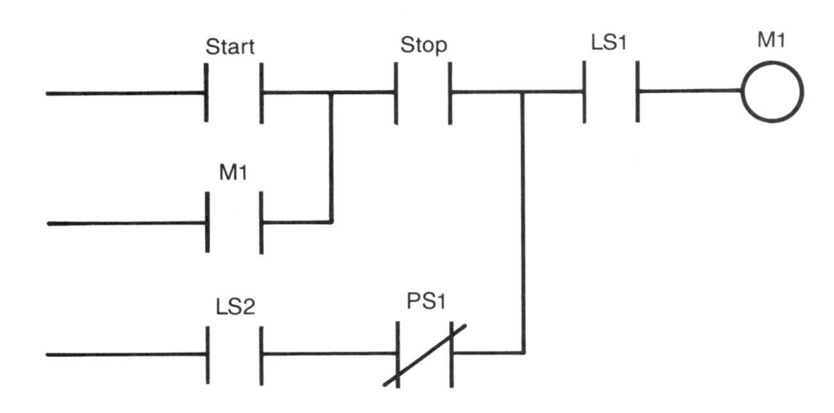

Figure 8-1. Problem 8-8.

8-9 Draw the equivalent contact symbology diagram for the logic circuitry shown in Fig. 8-1.

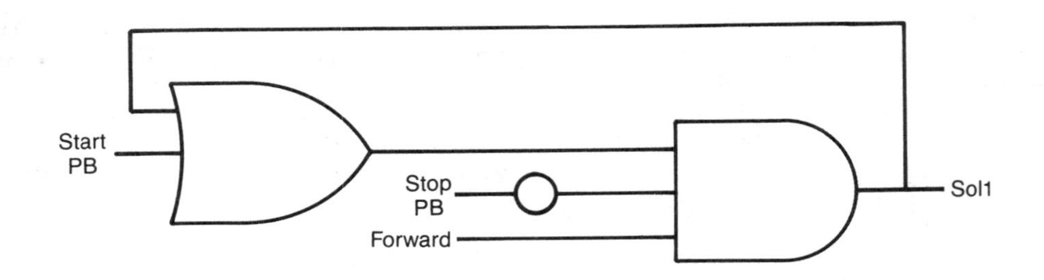

Figure 8-1. Problem 8-9.

8-10 Draw the equivalent contact symbology diagram for the circuit shown in Fig. 8-2.

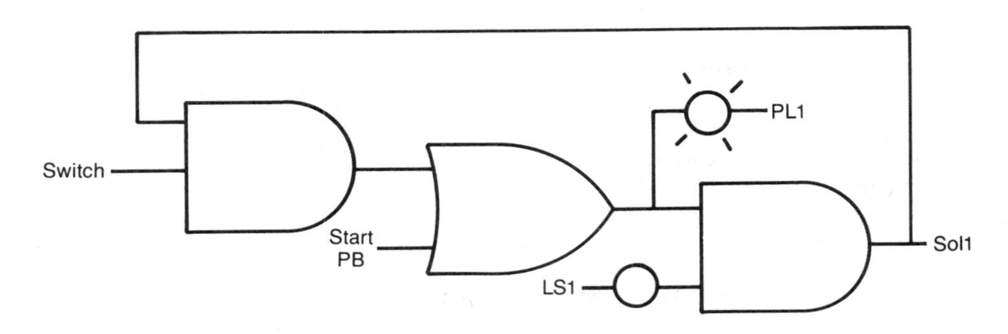

Figure 8-2. Problem 8-10.

8-11 Describe the address assignment process and why it is required.

8-12 True or False. The address assignment only takes into consideration real inputs and outputs of the system. If false, explain why?

8-13 The numbering scheme associated with the I/O address assignment depends on ______________________.

8-14 I/O address assignments are typically represented in one of three number systems: ______________________, ______________________, or ______________________.

8-15 Assign the following inputs and outputs an address such that all inputs are grouped together and all outputs are grouped together. Assume modularity of two points per module and each module capable of handling eight points. Indicate the slot.

Inputs: Limit switch LS10
Pressure switch PS3
Motor contact M11
Start push button SPB1
Reset push button RPB3

Outputs: Solenoid Sol 3
Motor M11
Pilot light PL4

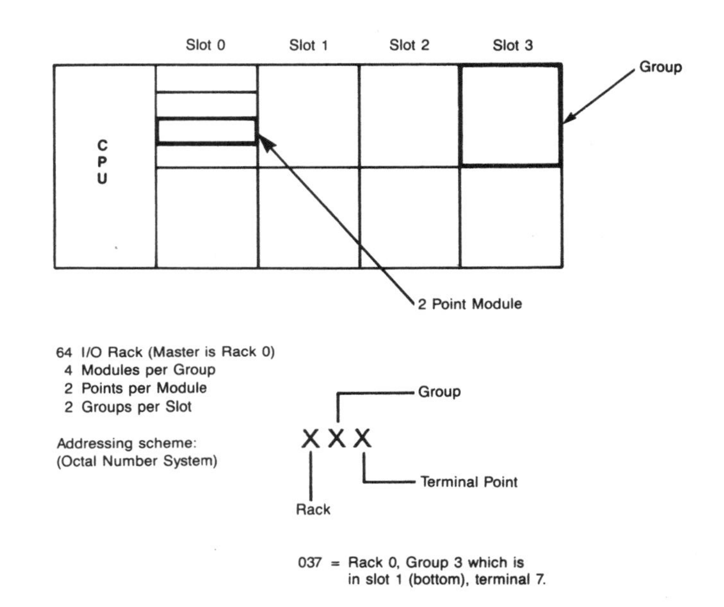

I/O Address Rack Group Terminal	Slot	Module Type	Description

Use this information for Problem 8-15.

8-16 Given a multiple contact input device, (e.g.,push button with 2 NO and 2 NC contacts), how many input points must be wired to the PC?

8-17 Using the circuit shown in Fig. 8-3:

a- Circle all real inputs and outputs
b- Assign the I/O addresses
c- Draw the I/O connection diagrams

Assume that the PC has modularity of 8 points per module and there are eight modules per rack; the master rack is numbered 0; the number system is octal.

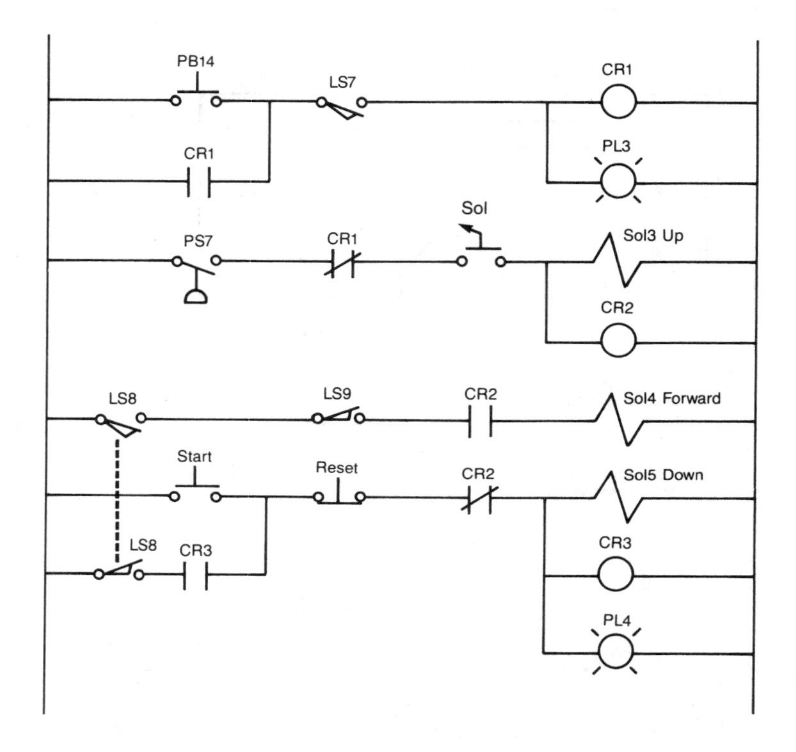

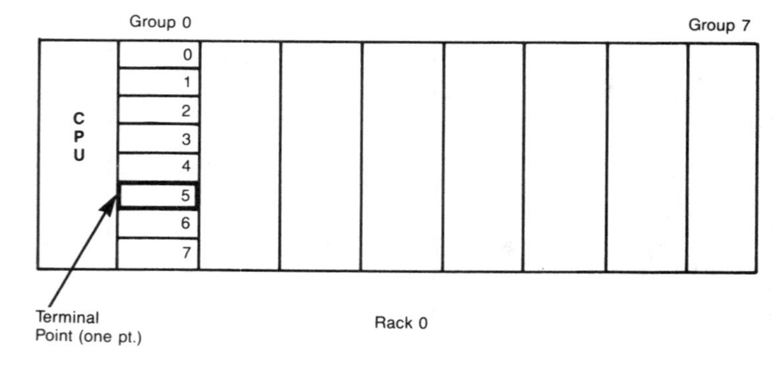

Figure 8-3. Problem 8-17.

8-18 In the circuit shown in Fig 8-4:

a- Identify real I/O by circling each
b- Assign the I/O addresses
c- Assign the internal addresses
d- Draw the I/O connection diagrams

Use the PC information of problem 8-17; internals start at 1000.

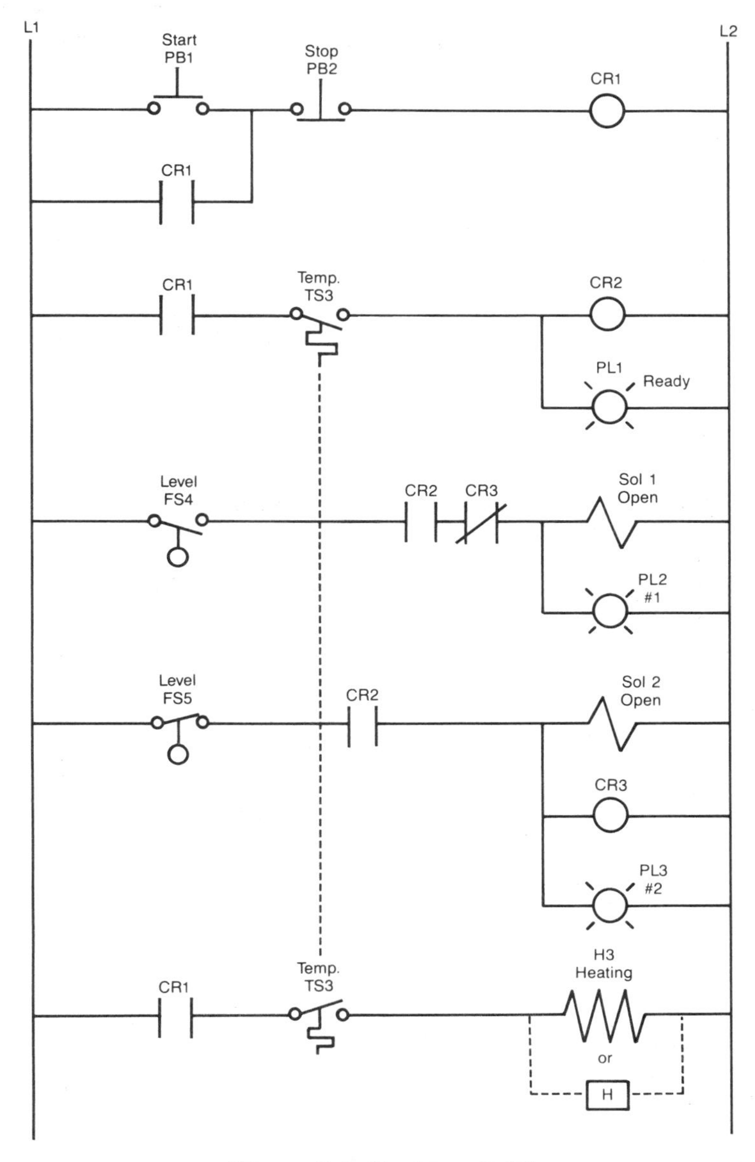

Figure 8-4. Problem 8-18.

I/O Address				
Rack	Group	Terminal	Module Type	Description
0	0	0		
0	0	1		
0	0	2		
0	0	3		
0	0	4		
0	0	5		
0	0	6		
0	0	7		
0	1	0		
0	1	1		
	.			
	.			
	.			
0	1	7		
	.			
	.			
	.			
0	7	7		

Use this information for part b of Problem 8-18.

8-19 The principal reason for leaving certain portions of the control circuit hardwired is to:

a- Minimize wiring
b- Avoid failure of main magnetic elements
c- Ensure safety
d- Keep some devices running at all times

8-20 The PC fault contacts are generally wired to other hardwired emergency circuit elements:

a- In parallel
b- In series
c- Normally open
d- Normally closed

8-21 The main reason the PC fault contacts are included in the hardwired circuit is:

a- To prevent system shut down
b- To detect I/O failures
c- To include the PC as an emergency stop condition
d- To shut down the system if there is a PC failure

8-22 What is improperly connected in the circuit shown in Fig. 8-5.

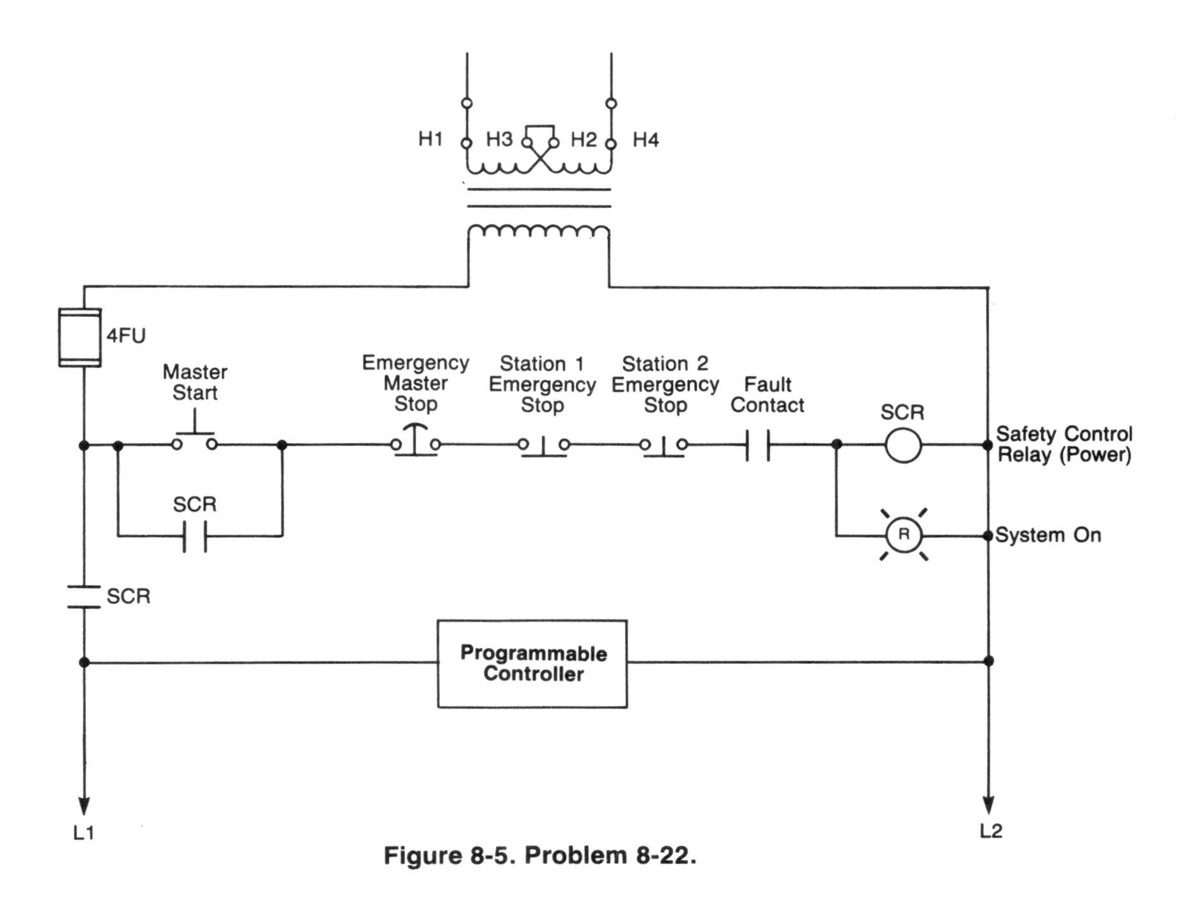

Figure 8-5. Problem 8-22.

8-23 In the circuit shown in Fig. 8-6, what is the purpose of the normally closed PC fault contacts? Describe what happens if the PC fails.

8-24 How would the PC fault contacts be incorporated in the emergency stop circuit if they are NO rather than NC?

8-25 Program ____________________________ is the process of translating logic or relay contact diagrams into PC ladder form.

8-26 Fill in the blanks with the appropriate device answer for the circuit shown in Fig. 8-7. The "X" indicates a real I/O as connected to the I/O module.

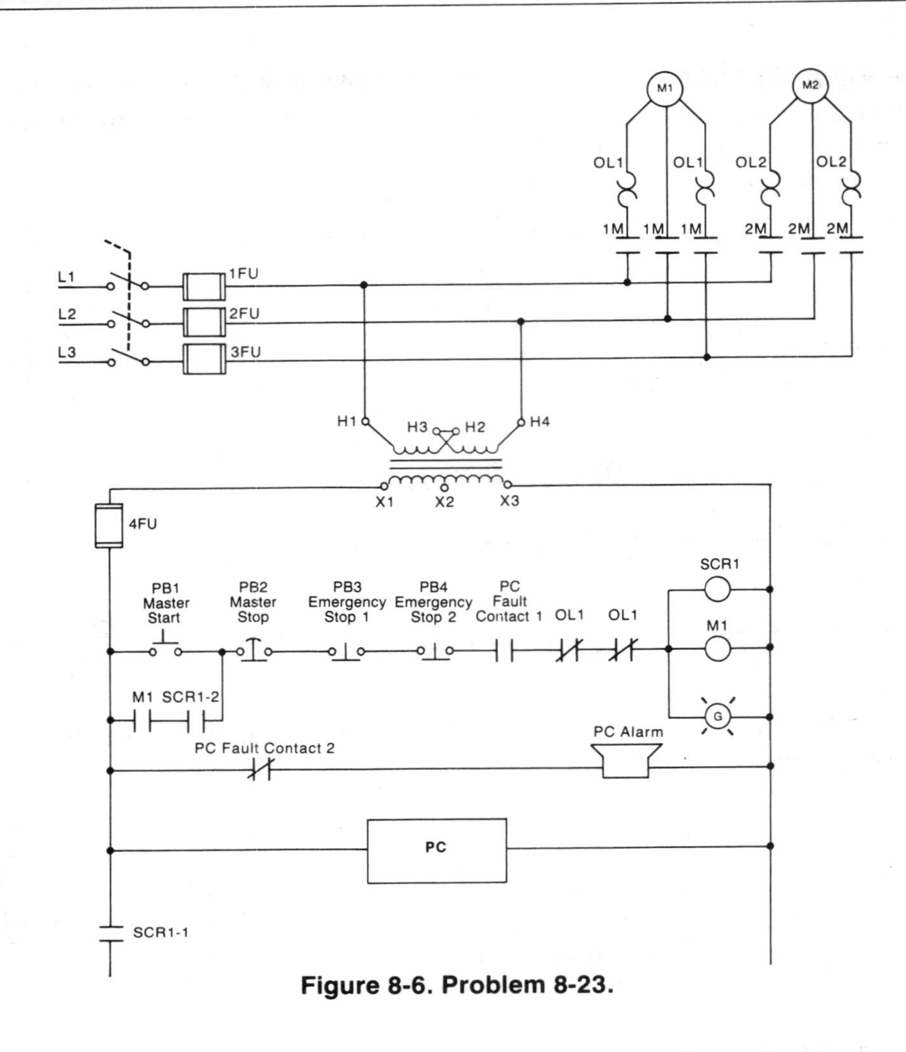

Figure 8-6. Problem 8-23.

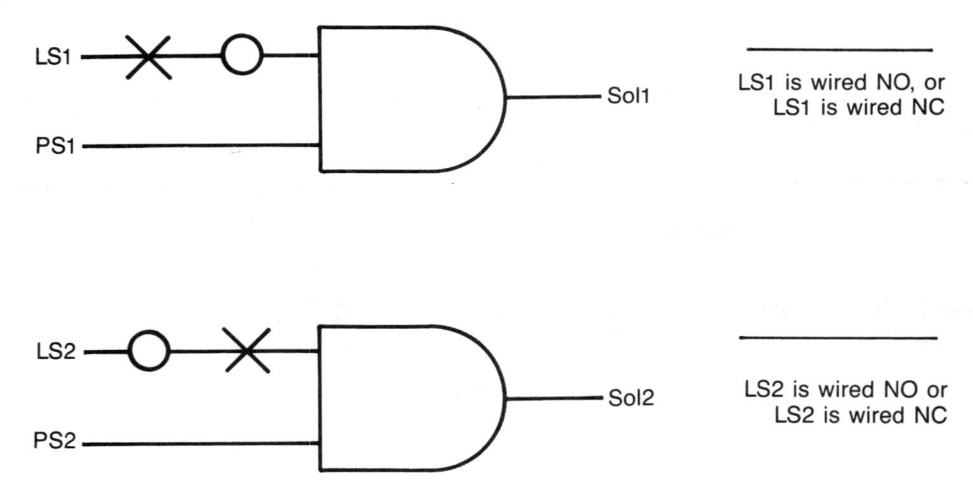

Figure 8-7. Problem 8-26.

8-27 Using the logic gate circuit shown in Fig. 8-8, generate the PC ladder program. Assume that internal output coils start at location 100 octal. For inputs, use addresses starting at 10; for outputs, use addresses starting at 50.

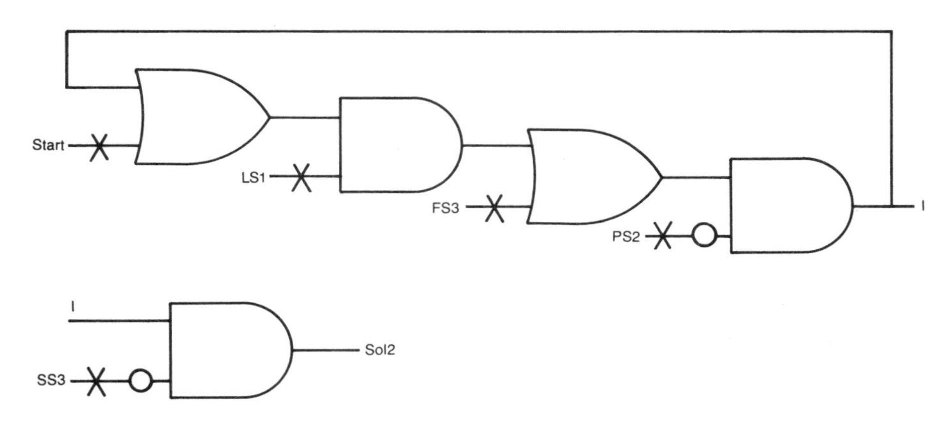

Figure 8-8. Problem 8-27.

8-28 Derive the PC ladder program from the relay circuit shown in Fig. 8-9. Use the PC information of problem 8-27.

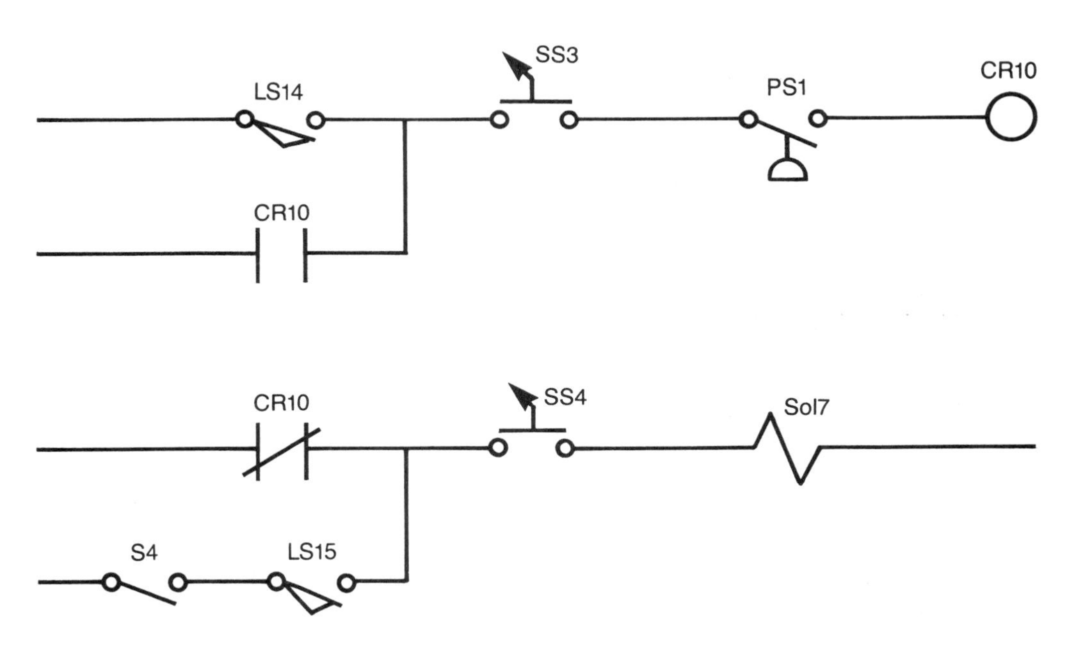

Figure 8-9. Problem 8-28.

8-29 True/False. Normally closed input devices are always programmed normally open.

8-30 Match each of the following with the appropriate answer(s):

____ Programmed NO	a- PC examines for OFF condition
____ NO contacts close	b- Input is ON
____ Programmed NC	c- PC examines for ON condition
____ NO contacts open	d- Input is OFF
____ NO contacts close	
____ NO contacts open	

8-31 What happens to the state of an input wired NO if it is programmed NC and the input is ON?

a- The input stays in the same state.
b- The contacts open.
c- The contacts close.
d- None of the above.

8-32 What happens to the state of a pushbutton input wired NC if it is programmed NC and the PB is depressed?

a- The input stays in the programmed state
b- The contacts are open
c- The module is energized
d- None of the above

8-33 True/False. A normally closed input field device is generally wired to the PC module in the same manner for fail-safe reasons.

8-34 True/False. An input field device is programmed NO or NC depending on how it is to operate. How the device is wired to the input module does not matter (NO or NC wired).

8-35 Write the PC ladder logic to implement the hardwired circuit shown in Fig.8-10 (for a and b). Show the I/O connection diagram.

a- Wire the NO limit switch to PC module.
b- Wire the NC limit switch to PC module.

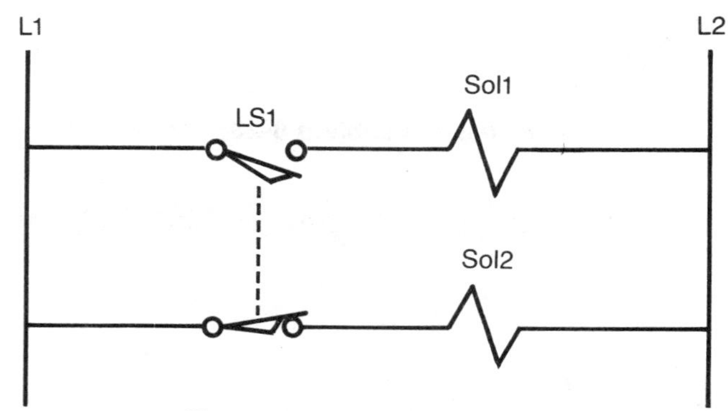

Figure 8-10. Problem 8-35.

8-36 In an application, a limit switch is wired NC to the input module and is programmed NO in order to act as a NC contact. During trouble-shooting of a failure, it is found that the limit switch is defective and needs to be replaced; however, the only available spare limit switch at the time is a NO. What would need to be changed in the program?

8-37 Name three reasons why some machines are modernized.

8-38 Using the following specifications, implement the PC program including I/O connection diagrams for the circuit shown in Fig. 8-11.

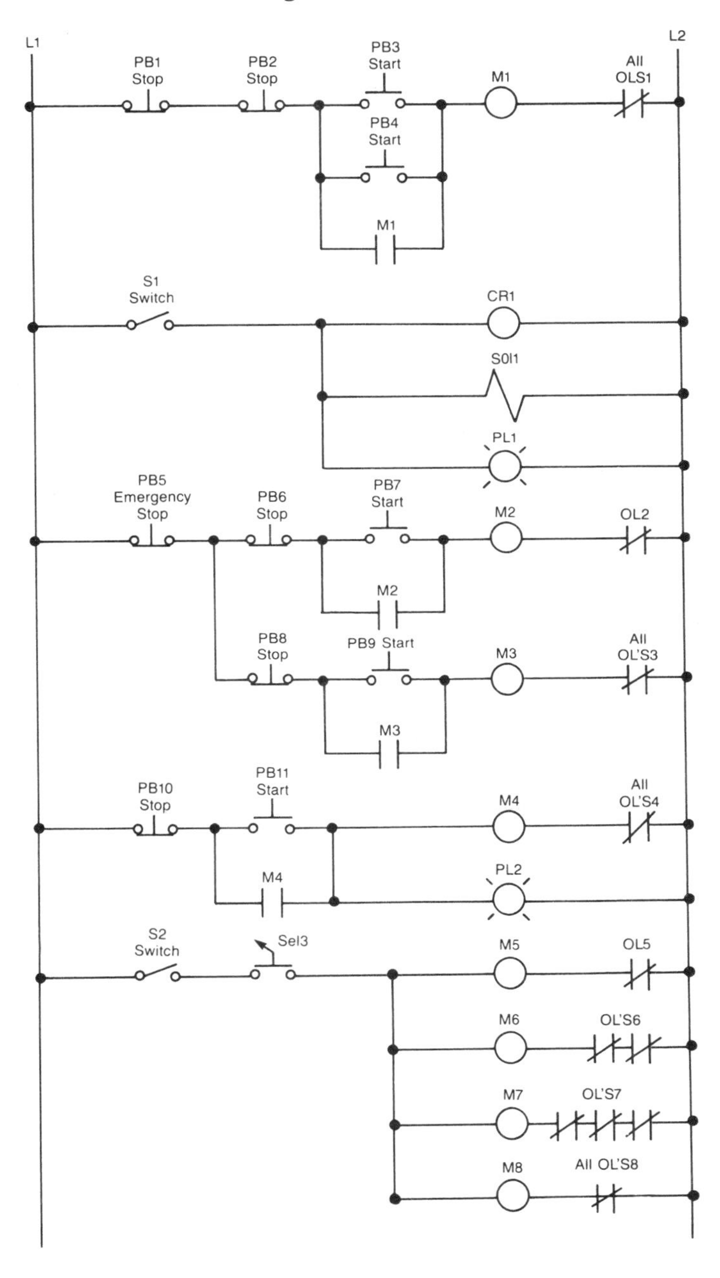

Figure 8-11. Problem 8-38.

Continued. Problem 8-38.

a- All stops are wired NC.
b- All stops are changed to NO but operate as NC.
c- Auxiliary contacts M1 and M4 are brought as inputs, and overloads OL2 and OL5 are also inputs.

Note: Use input addresses 0 to 27, and output addresses 30 to 47; internals start at 100 and timers at 200. Use the octal number system; show the addresses in the I/O connection diagram.

8-39 Implement the PC ladder diagram in Fig.8-12. The stop pushbutton is to be wired using the NO connection. Show the I/O addressing and connecting diagrams.

8-40 Implement the circuit specified in problem 8-39, with additional inputs from the forward overload (one) and the reverse overload (one), and contacts M1 and M2.

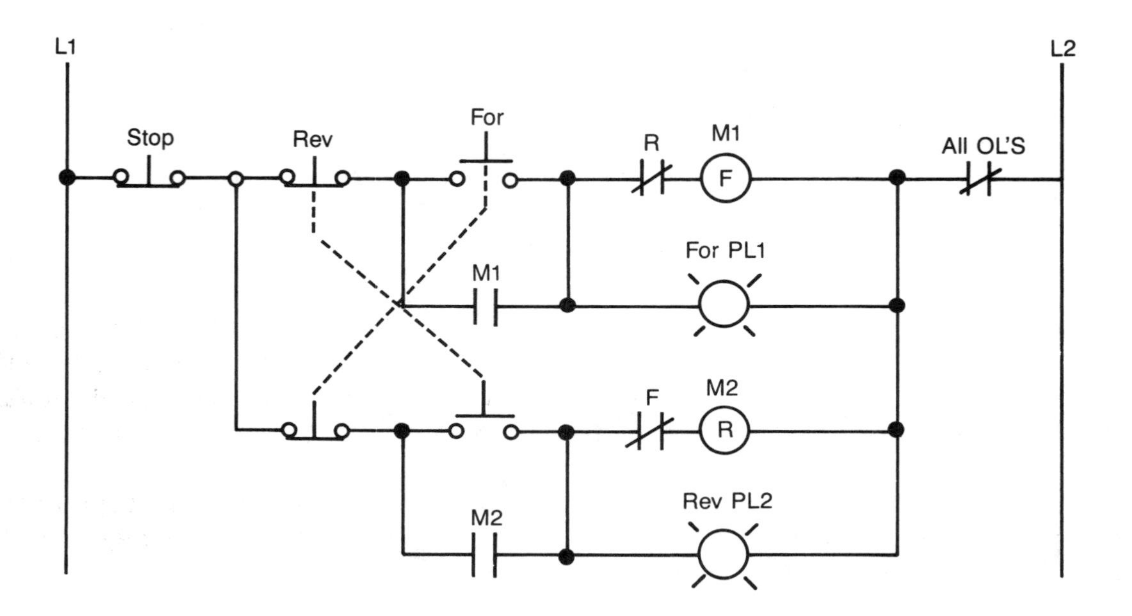

Figure 8-12. Problem 8-39.

8-41 Implement the circuit shown in Fig.8-13 in PC ladder logic. Include I/O address assignments and connections. (Hint: Use an internal output to trap or interlock the instantaneous timer contacts).

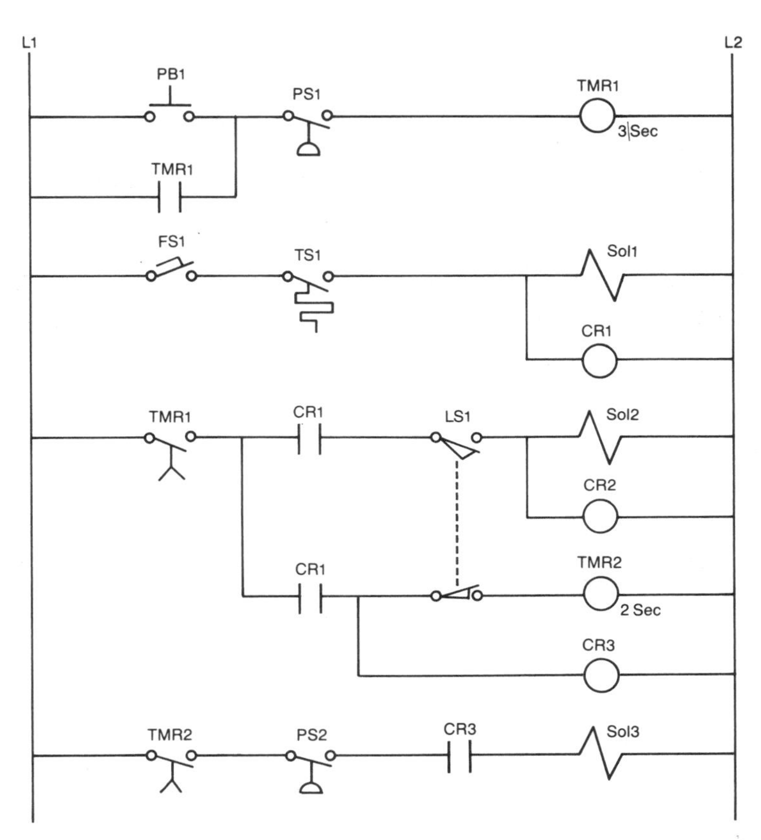

Figure 8-13. Problem 8-41.

8-42 A variable speed AC drive is connected to an operator's manual station as shown in Fig.8-14. You are required to interface the control signals START/STOP, JOG/RUN, FORWARD/REVERSE, and analog output to the drive and implement the START/STOP, JOG/RUN, FOWARD/REVERSE logic in a PC. (Use the PC specifications of problem 8-38).

(Hint: Note that the START/STOP common is not the same as the controller's common, and the FORWARD/REVERSE switches the controller's common. You may use 115 VAC and dry contact output modules).

8-43 For the previous problem, add a selector switch for an AUTO/MANUAL operation so that the drive can operate under the PC control or the operator station. Include the AUTO/MANUAL input signal to the PC for indication.

8-44 For the system shown in Fig.8-15, implement a control program that will detect the position of the bottle, wait 0.5 seconds, and then fill the bottle until the photo sensor detects the filled position. After the bottle is filled, wait 0.5 seconds to continue to the next bottle. Include the start and stop circuits for the outfeed motor and/or the start of process.

Note: Use same PC specifications of problem 8-38.

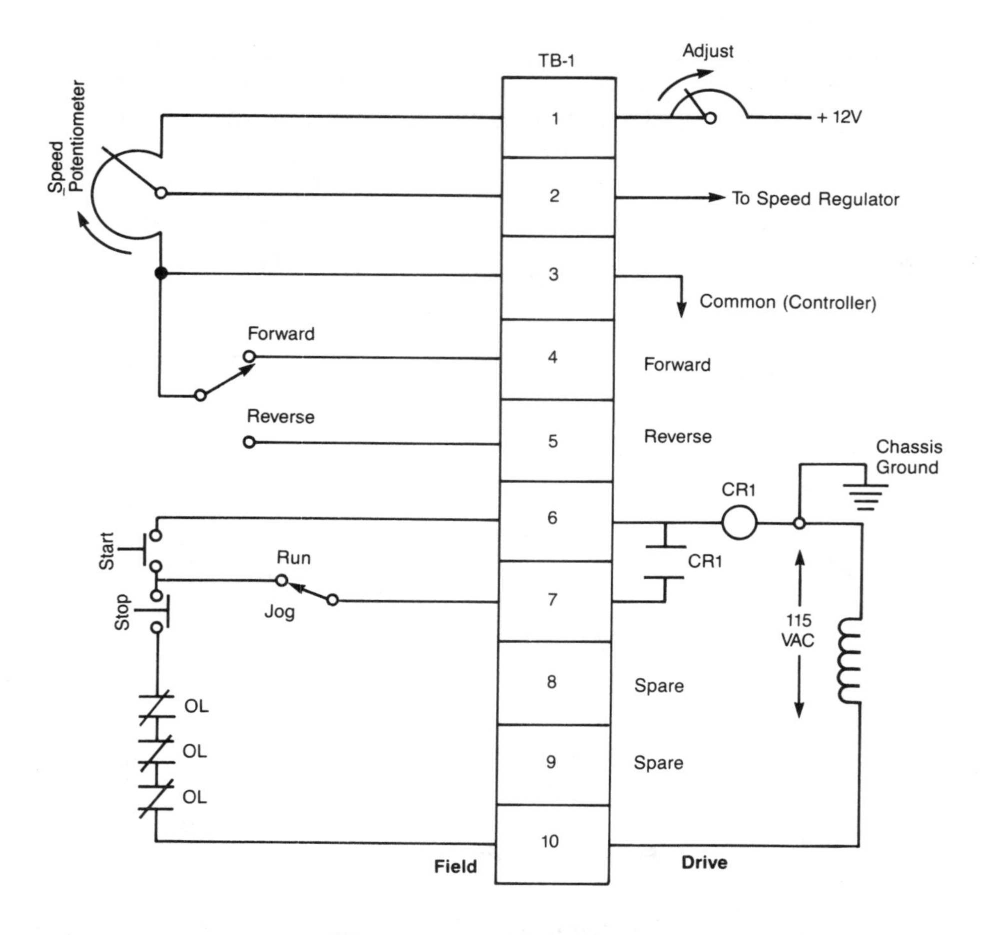

Figure 8-14. Problem 8-42.

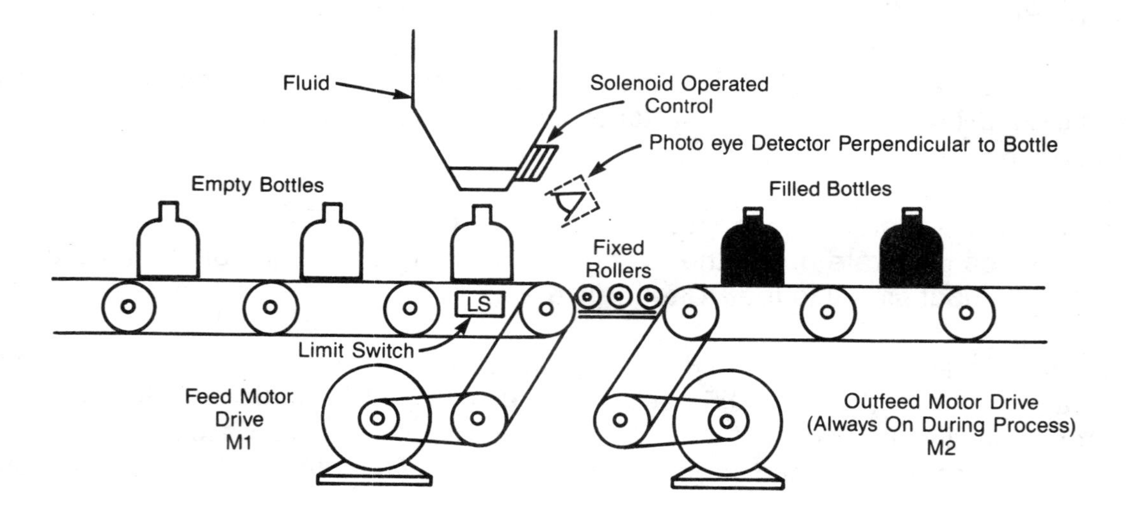

Figure 8-15. Problem 8-44.

8-45 Implement the electromechanical control section of the circuit shown in Fig.8-16. Include a PC fault contact (NO). The PC does not have enable timer inputs; therefore, trap the timer contacts using internals. The PC system has capacity for 512 I/O (000 to 777 octal); start inputs at 000 and outputs at 030. MCR addresses are from 2000 to 2037 and internals from 1000 through 1777; timers start at 2040 through 2137. Show the portions to leave hardwired, the real I/O address assignment, the internal address assignment, and the PC ladder program.

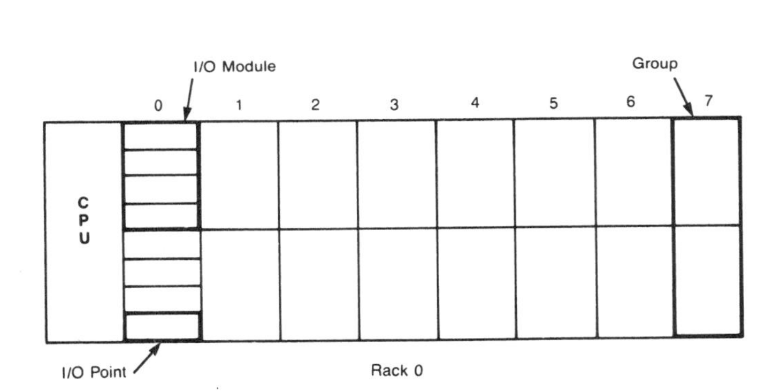

Block schematic for the PC used in problem 8-45. Each rack can have 64 I/O, modularity in 4 points per module, 8 points per group.

Example:

Rack	Group	Terminal Point	
0	0	0	First I/O (terminal point 0 in group one)
0	1	3	Last I/O in first I/O module in group one (terminal point 3)

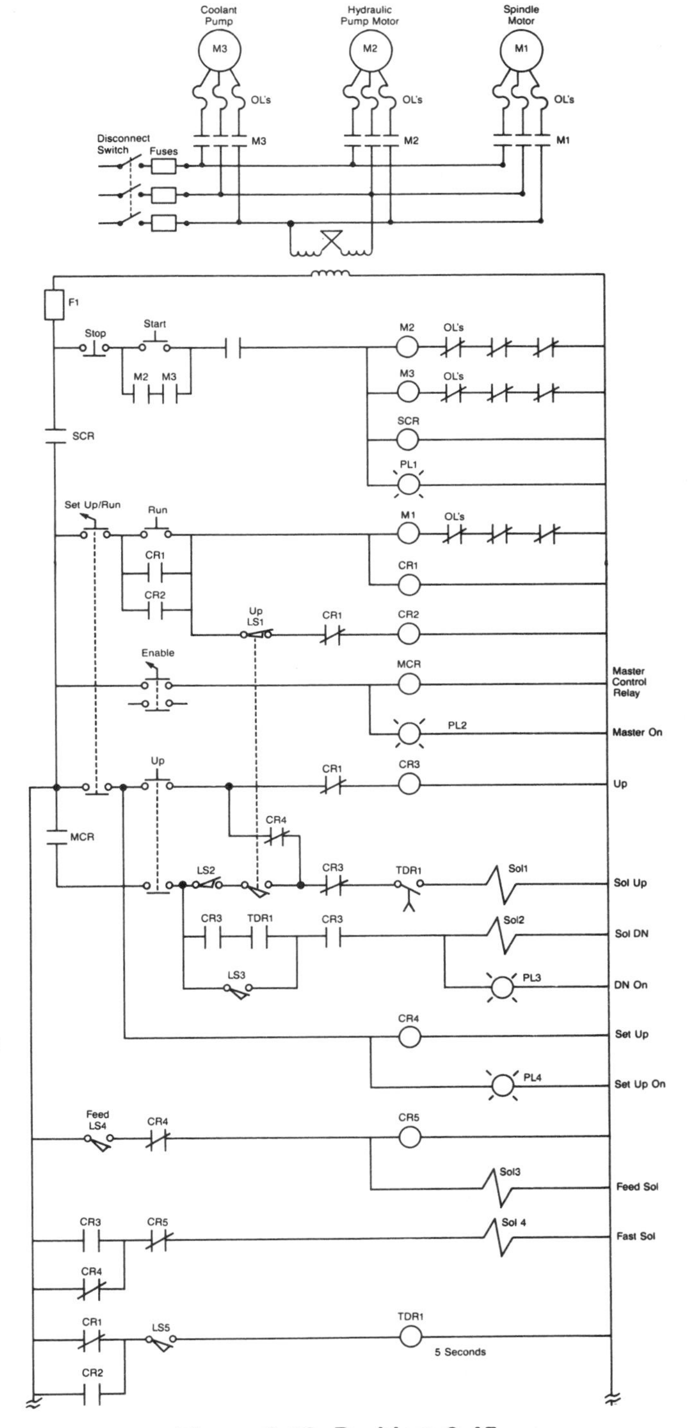

Figure 8-16. Problem 8-45.

8-46 Figure 8-17 illustrates a flow diagram of a batch/mix application. It is required to control the sequence of events and measurements.

Two ingredients, A and B, are to be mixed, and a cleanser is used to clean the ingredients line. A pump motor provides the necessary pressure to send the ingredients through the lines; a command initiates pump forward flow and another for reverse flow.

The procedure is as follows:

1- The pump motor is always ON and is activated by a FP (Forward Pump signal 115 VAC). Ingredient A is sent to the tank first, FP ON and SOL1 ON; the flow meter M gives one pulse for every gallon of flow. Solenoid valve 4 must be open (ON) to let A into the reactor tank (SOL5 OFF). Solenoid valves 1 and 4 will be open until 500 gallons have poured in.

2- After ingredient A is in the tank, the line should be cleaned. Before cleaning, however, the remaining ingredient A in the line should be pumped back into the A tank by energizing the RP (reverse pump). Wait 0.8 seconds for change between FP and RP; SOL4 should be closed. The amount of ingredient A that is left in the line has been calculated to be approximately 83 gallons. If the meter does not give any pulses after 5 seconds, it should be assumed that there is no more ingredient A in the line.

3- After the line has been cleaned of ingredient A, SOL3 should be open to let the cleanser go through, solenoids 1 and 4 OFF and SOL5 ON. The amount of cleanser should be 90 gallons as detected by the meter. After the gallon amount of cleanser has passed, SOL3 should be closed and SOL7 should open to let air flow for 5 seconds.

4- After cleaning the line, ingredient B should be added (330 gal); the process of adding follows the same procedure as ingredient A.

5- After B is in the tank, the mixer motor should start for 5 minutes while reversing the pump to get the rest of ingredient B in the line.

6- After the mixing is completed, SOL6 should open and let the mixed batch go into a finished tank.

7- (Optional, not in answer) The reactor tank should be flushed by pouring the cleanser to the tank until the float switch, FS1, is ON indicating the top of tank. Mix for 20 seconds and release the cleanser through SOL6 until the lower float switch is deenergized (empty).

You are required to:

a-Implement a flow chart diagram of the process
b-Implement the logic using logic gates, contact symbology, or both
c-Implement the I/O addressing and I/O connection diagram
d-Implement the PC ladder diagram (you may use generic blocks if required)

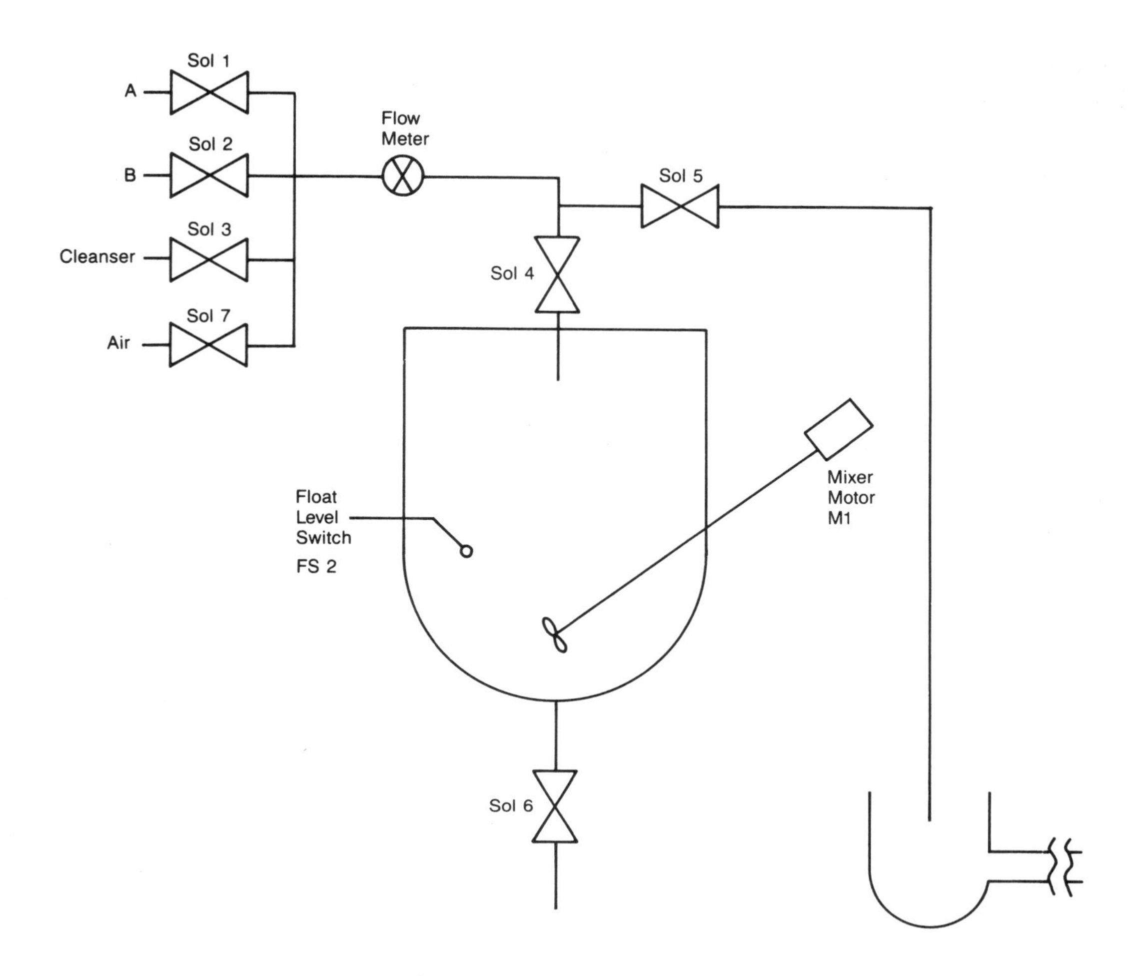

Figure 8-17. Problem 8-46.

8-47 The exclusive-OR circuit (shown in Fig. 8-18) will energize an output if and only if one of the inputs is ON at a time. The exclusive-NOR function will operate an output if and only if both of the inputs are ON or both are OFF at a time. This function can be implemented using an internal and a NC contact of the internal in series with the output (for a NOR) as shown in Fig. 8-18. Implement the exclusive-NOR without using an extra output coil.

8-48 True/False. A transitional or one-shot output can be ON for less than one scan.

8-49 True/False. In the circuit shown in Fig. 8-19, internal 1 will never turn ON. Explain.

8-50 Draw the timing diagram for the oscillator circuit shown in Fig. 8-20.

8-51 True/False. If timer T2 is programmed before T1 for the circuit shown in Fig. 8-20, the flash function will still operate.

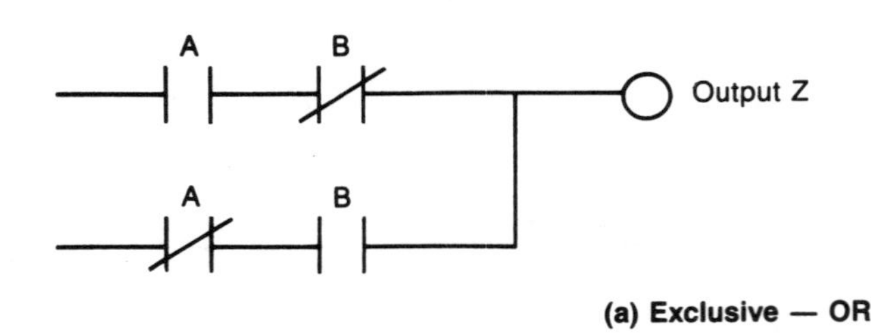

Truth Table		
A	B	Z
0	0	0
0	1	1
1	0	1
1	1	0

(a) Exclusive — OR

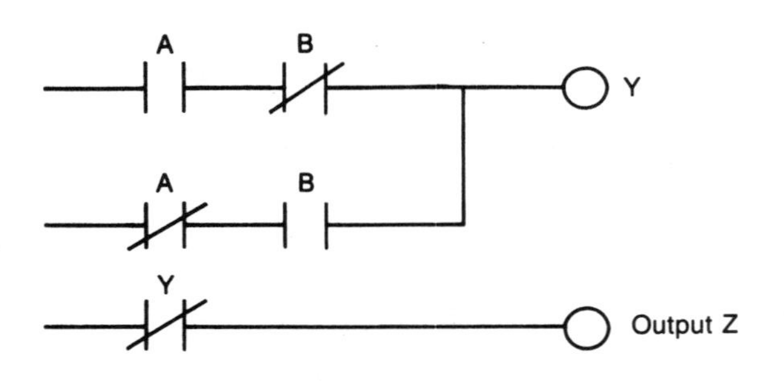

Truth Table		
A	B	Y
0	0	1
0	1	0
1	0	0
1	1	1

(b) Exclusive — NOR

Figure 8-18. Problem 8-47.

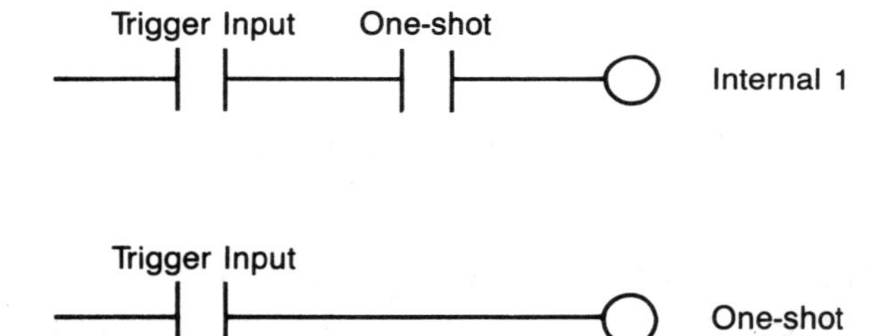

Figure 8-19. Problem 8-49.

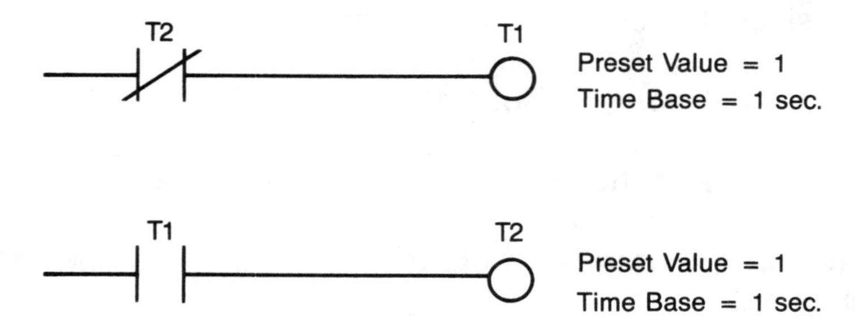

Figure 8-20. Problem 8-50.

8-52 The one second ON pulse used in the flashing circuit comes from T2 and can be seen on a CRT monitor changing states every second. The output of T1 would not be seen under a monitor mode in the CRT. Why not?

8-53 Implement a "trap" circuit that will detect the output of timer T1 for Fig. 8-20.

8-54 Draw the timing diagram for the circuit in Fig.8-21 (for preset value PV of 1 sec and 2 sec) Indicate the time between pulses.

Figure 8-21. Problem 8-54.

8-55 The oscillator circuit of Fig 8-21 can be used to generate a one second pulse for a clock. If the scan time of the PC is 20 msec and the time base of the timer is one hundredth of a second, what should the preset value be for a one second cycle at the OFF-to-ON transition of timer contacts:

a- PV = 98
b- PV = 96
c- PV = 96 for accuracy after first 1 sec cycle
d- PV = 94 for accuracy after first 1 sec cycle

8-56 The maximum count for a counter in some PCs is 32767. Using the instructions and specifications in Fig. 8-22, write a program that would allow a 6 digit, seven-segment display count to 999999.

Use registers starting at R100. The slot location for the first four LED digits is 3, and the slot for the other two is 4. The event count LS1 (end of cycle) is connected to input 17.

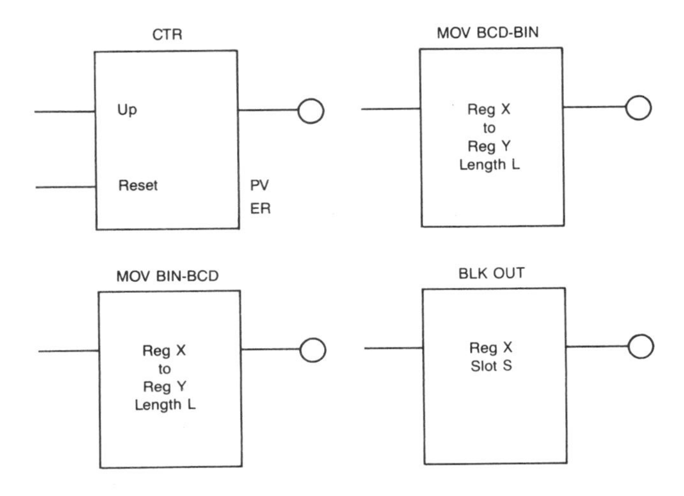

Figure 8-22. Problem 8-56.

8-57 Using as few steps as possible implement the circuit in Fig.8-23 in PC ladder form, eliminating any possible bidirectional power flow. (Each ladder rung solution has 10 contacts).

Hint: Start at marked nodes.

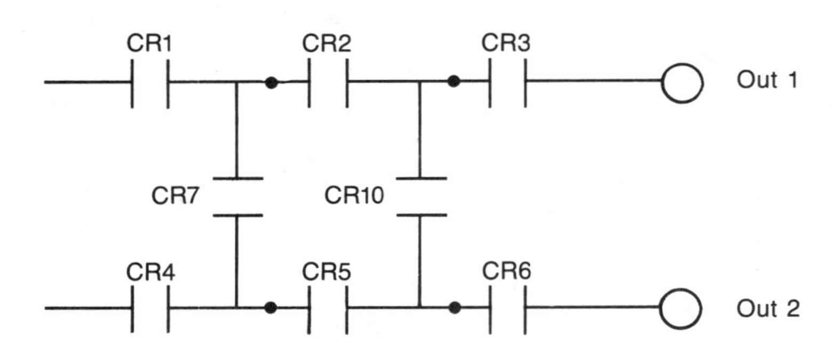

Figure 8-23. Problem 8-57.

8-58 Draw the timing diagram for the push-to-start/push-to-stop circuit shown in Fig. 8-24.

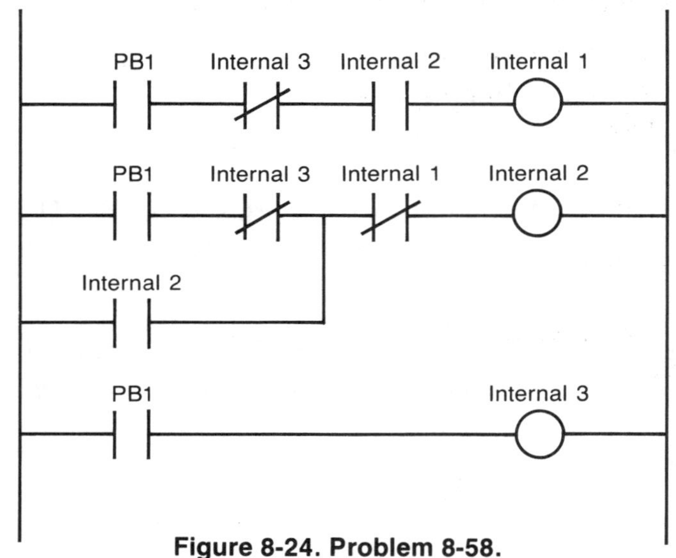

Figure 8-24. Problem 8-58.

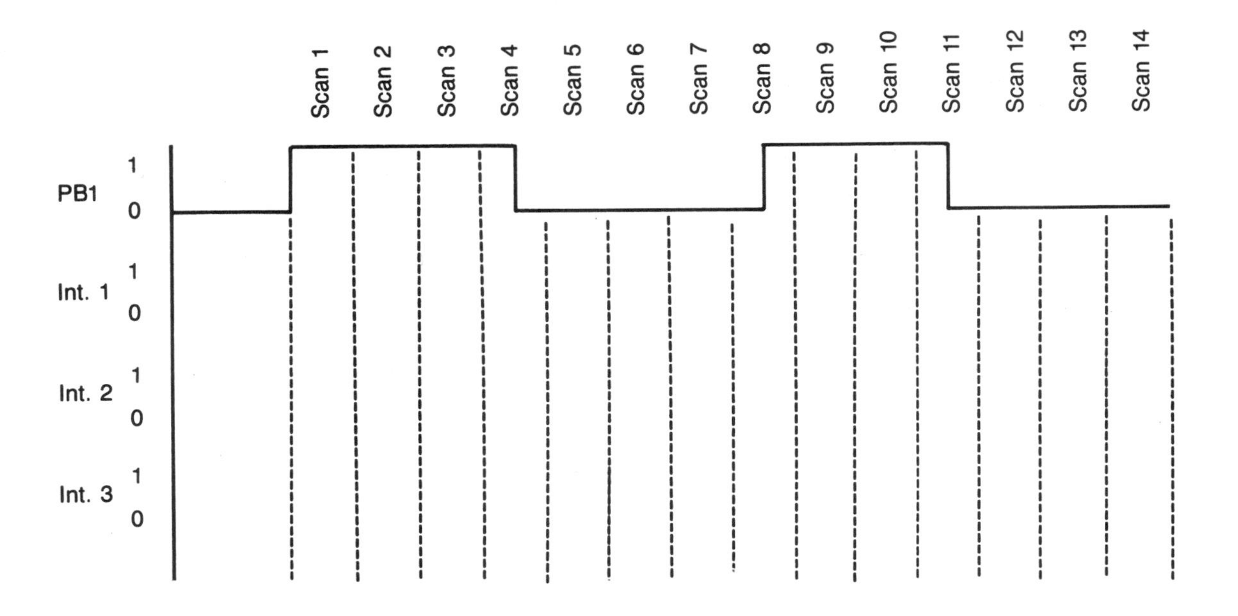

Figure 8-24 Continued. Problem 8-58.

8-59 What is the function of internal 1 in the push-to-start/ push-to-stop circuit in Fig.8-24:

a- Detection of first push of PB1
b- Detection of second push of PB1
c- One shot pulse
d- a and b
e- a and c
f- b and c
g- All of the above

8-60 True/False. The push-to-start/push-to-stop circuit in Fig.8-24 would work if the logic rungs are interchanged.

UNIT

--]9[--

DOCUMENTING THE SYSTEM

- Documentation can be defined as an orderly collection of recorded information, concerning both the operation of a machine or process and the hardware and software components of the control system.
- The achievement of proper documentation is realized through the gathering of hardware as well as software information.
- A good documentation package must include a system abstract, system configuration, I/O wiring connection diagram, I/O address assignment, internal I/O address assignment, register assignment, program coding printout, and reproducible stored programs.
- The system abstract should provide a clear statement of the control task, a description of the design strategy or philosophy used to implement the solution of the problem, and a statement of the objectives that must be achieved.
- The system configuration is a system arrangement diagram that pictorially defines the location, simplified connections, and minimum details of the hardware components as defined in the system abstract.
- The I/O wiring diagram shows the actual connections of the field input and output devices to the PC modules.
- The I/O address assignment identifies each field device by an address, the type of input or output modules, and the function this device performs in the field.
- Internal I/O address assignments and register assignments are an important part of the total documentation because they specify the internal coils used for interlocking and the register usage and description. This documentation section will avoid the use of already defined registers and internals.
- The program coding printout is a hard copy of the control program stored in the PC memory. This printout is the ultimate program back-up if it is updated regularly.
- The PC program is generally stored in a reproducible type of device such as cassette tape, floppy disk, microdisk, or electronic memory module. These types of storage devices are useful and necessary since the CPU of the system used for controlling the process is usually located remotely from where the program is being developed; therefore, a program that has been stored in a reproducible device may be loaded in the controlling PC.
- Documentation systems provide an alternative to manual documentation, reducing manpower and turn-around drafting time. The following are features that are provided in documentation systems:

- Program titles
- Multiple subtitles
- Date and time the documentation was last produced
- Page numbering
- Extensive commentaries before and after each rung
- Contact or element description
- PC address for each contact
- Pictorial representation of each PC instruction (coils, contacts, etc.)
- Rung numbers
- Rungs where each contact is used
- All preset values of registers used
- Identification of internals and real I/O

REVIEW QUESTIONS

9-1 The ______________________ package can be defined as an orderly collection of recorded information that concerns not only the software and hardware components of a system, but also the operation of the machine or process control.

9-2 True/False. The documentation of a control system should start in the final design stage.

9-3 True/False. Proper documentation is only realized through an ongoing process of recording descriptive details concerning the application software.

9-4 Which of the following is not part of the documentation package:

a- System configuration
b- System abstract
c- I/O address assignment
d- Usage of ladder instructions
e- I/O wiring connection diagram
f- Internal I/O address assignment
g- Register assignment
h- Reproducible stored programs
i- Program printout

9-5 A programmable controller system has a main CPU connected to a minicomputer through a RS-232C port. Both are located in the control room. A CRT display and printer are also connected to the main CPU. The main unit has 16 inputs and 16 outputs (115 VAC). The main processor communicates with its three subsystem racks, configured in a star configuration, through a serial interface module located in each subsystem.

Rack #1 (128 I/O) is located in the warehouse and handles 30 inputs and 58 outputs; the I/O addresses for this rack are at 200 through 377.

Rack #2 (128 I/O) is controlling a paper machine with 35 inputs and 76 outputs with addresses 400 through 577.

Rack #3 (64 I/O)and rack #2 are located on the main plant floor; rack #3 controls a winding machine. The winding machine has 20 inputs and 30 outputs with addresses 600 through 677.

The addresses are all octal, all I/O is 115 VAC, and the I/O receiver must be located in the first slot of each remote rack. The modularity is eight points per module. For this system, draw a system arrangement diagram.

9-6 Create the I/O address assignment sheet for the wiring connection diagram shown in Fig. 9-1.

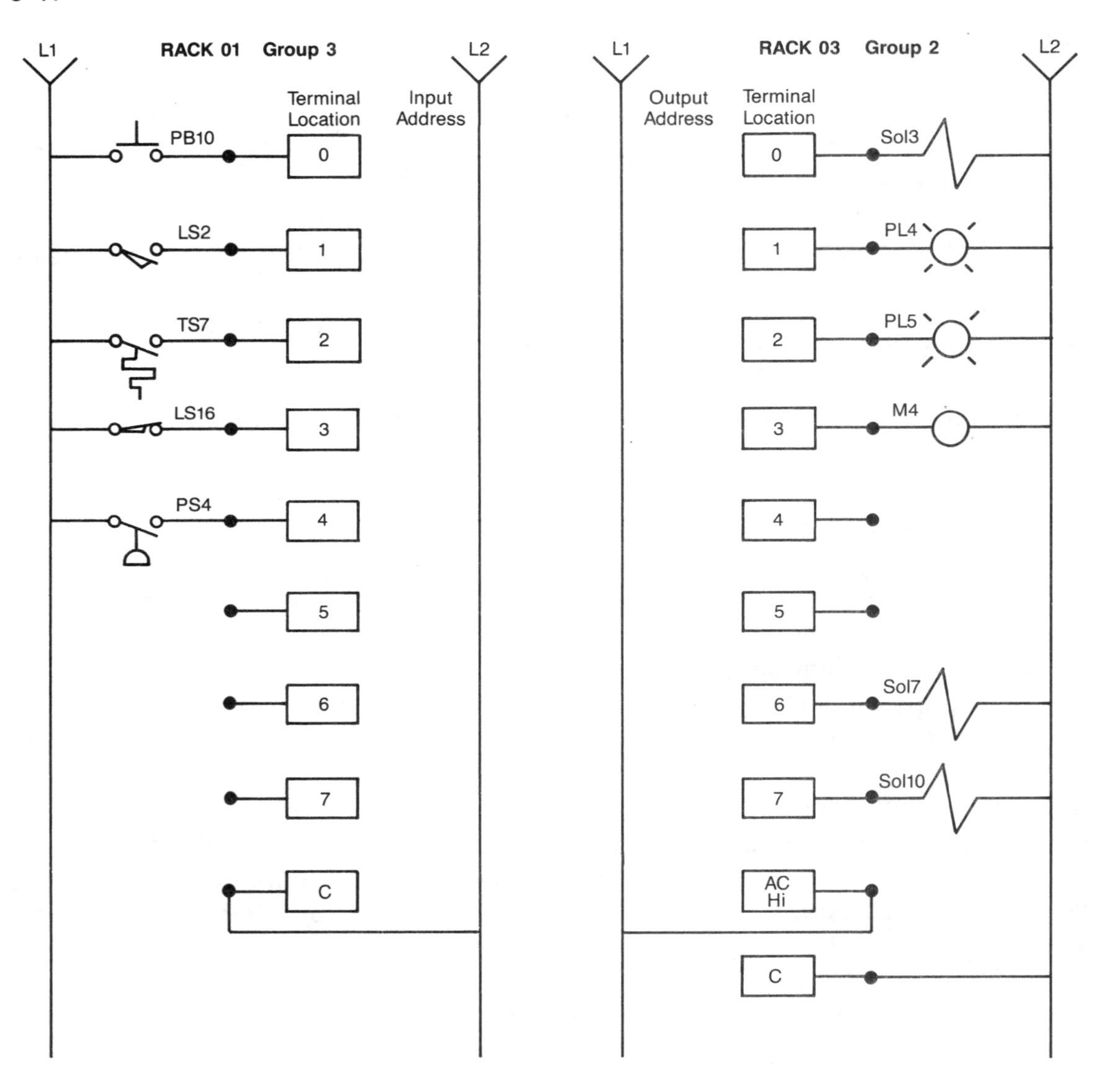

Figure 9-1. Problem 9-6.

9-7 A description of the design strategy should be included in the:

a- System configuration
b- System abstract
c- Register assignment

9-8 True/False. A full understanding of the control system is possible by having only the program printout.

9-9 Which of the following documents show actual connections of field inputs/outputs?

a- I/O address assignments
b- Register assignments
c- I/O wiring connection diagrams
d- System configuration

9-10 True/False. System documentation will be helpful only during installation and start-up.

9-11 As part of the register assignment, you should include the registers that are:

a- Used
b- Not used
c- a and b

9-12 What information should be included on the I/O address assignments documents?

9-13 True/False. Wire color changes should be documented.

9-14 The assignment of internal outputs should be done during:

a- The assignment of real I/O
b- After the assignment of real I/O
c- As the internals are used during development
d- During the I/O wiring connections

9-15 The register assignments, when properly documented, will prevent:

a- Multiple use of registers in the program
b- Improper reference of a register
c- Cross-reference of register usage
d- Register changes and deletions

9-16 An up-to-date ladder printout will not provide:

a- A replica of what is in memory
b- A hard copy of the actual control program in the PC
c- The I/O connections of field devices
d- The latest software revision of the program

9-17 Name three mediums that are used to store reproducible programs.

9-18 The safest program back-up source is:

a- Reproducible storage devices
b- Latest ladder printout
c- The program in the PC RAM memory
d- The I/O wiring and address assignment document

9-19 Documentation systems generally provide:

a- Reduction of drafting manpower
b- Explanations of ladder rungs
c- Extensive program listings
d- All of the above

9-20 Name eight features that are commonly available in documentation systems.

UNIT

--]10[--

INSTALLATION, START-UP, AND MAINTENANCE

SYSTEM LAYOUT

- The system layout is a conscientious approach in placing and interconnecting the components not only to satisfy the application, but also to insure that the controller will operate trouble free in the environment in which it will be placed.
- The system layout takes into consideration not only the PC system, but also other equipment such as isolation transformers, auxiliary power supplies, safety control relays, and incoming line noise suppressors.
- The best place for the PC enclosure to be located is near the machine or process that it will be controlling.
- Enclosures should conform to NEMA standards.
- The temperatures inside the enclosure should not exceed the maximum operating temperature of the controller, typically 60°C.
- If "hot spots" develop inside the enclosure in a system, a fan or blower should be installed; if condensation occurs, a thermostatically controlled heater may be installed.
- The system enclosure (with the PC) should not be placed close to equipment generating high noise such as welding machines.
- To allow maximum convection cooling, all controller components should be mounted in a vertical (upright) position.
- Grouping of common I/O modules is a good practice. All AC wiring should be kept away from low-level DC wiring to avoid cross-talk interference.
- The duct and wiring layout defines the physical location of wireways and the routing of field I/O signals, power, and controller interconnections within the enclosure.
- Proper grounding techniques specify that the grounding path must be permanent (not solder), continuous, and able to conduct safely the ground-fault current in the system with minimal impedance.
- Using isolation transformers is a good practice if noise is likely to be introduced in the power lines by noise- generating equipment.
- Constant voltage transformers should be used when there is a soft AC line.
- A capacitor is sometimes placed across the power disconnect to protect against outrush, which occurs when the output triacs are turned off by throwing the power disconnect, thus causing the energy stored in the inductive loads to seek the nearest path to ground, which is often through the triacs.

- Typical conditions specified by manufacturers are set when 60% of the inputs are ON at any one time, 30% of the outputs are ON at any one time, currents supplied by all modules average a certain value, and the ambient temperature is around 40°C.

INPUT/OUTPUT INSTALLATION

- Preliminary wiring considerations include the wire size, wire and terminal labeling, and wire bundling.
- When placing an I/O module, the type of module must be verified as well as the slot address or location as defined by the I/O address assignment.
- A bleeding resistor may be used in cases in which the output of a field device exhibits a current leakage that may turn ON the input circuitry. Output modules generally have a specified leakage current.
- Inductive loads should be suppressed using RC snubbers and/or MOVs.
- If fuses are not incorporated in the output modules, they should be installed externally (at the terminal block).
- Shielded cables are characterized by having at least one- inch lay, or approximately twelve twists per foot. These cables should be grounded at one end only, preferably at the chassis rack.

THE SYSTEM START-UP

- The system start-up includes pre-start-up procedures, static input wiring checks, static output wiring checks, pre-start-up program checks, and the dynamic checkout.
- The pre-start-up procedure involves several inspections of the hardware components before power is applied to the system.
- The static input wiring check is performed with power applied to the controller and input devices.
- The static output wiring check is performed with power applied to the controller and output devices. All the devices that will cause mechanical motion should be disconnected locally.
- The pre-start-up program check is simply a final review of the control program.
- The dynamic checkout assumes that all static checks have been performed, the wiring is correct, the hardware components are operational, and the software has been thoroughly reviewed. This checkout is a procedure by which the logic of the control program is verified for correct operation of the outputs according to the logic program.

MAINTENANCE AND TROUBLE-SHOOTING

- Even though a PC system requires minimal maintenance, a simple revision of air filters and a check for tight I/O modules and wiring may be highly desired. A periodic check should be done when the machine or equipment is scheduled for its preventive maintenance.
- As a rule of thumb, 10% of I/O modules and one of each main board should be kept as spare parts.
- When diagnosing I/O malfunctions, the first check should be the LED power and/or logic indicators in the module.
- When diagnosing I/O malfunctions, the best method is to isolate the problem either to the module or to the field wiring.

REVIEW QUESTIONS

10-1 Briefly describe what is termed the system layout.

10-2 True/False. With a proper system layout, the components would be easily maintained, but components may not be easily accessible.

10-3 True/False. The system layout takes into consideration other components that form the control system.

10-4 Name three types of equipment other than the PC that could form part of the system layout.

10-5 The best location for the PC enclosure is:

a- Close to the incoming power
b- In the control room
c- Close to the machine or process
d- Far away from the machine or process

10-6 Placing a remote I/O panel close to the controlled machine will generally:

a- Simplify start-up
b- Minimize wire runs
c- Simplify maintenance and trouble-shooting
d- All of the above

10-7 Metal NEMA enclosures are recommended since they provide protection from all of the following except:

a- Atmospheric contaminants
b- Vibration
c- Conductive dust
d- Moisture

10-8 Name four enclosure layout considerations for the placement of components, wiring of I/O, and location of the enclosure.

10-9 True/False. Placing AC power outlets inside the enclosure should be avoided when possible.

10-10 Typically, programmable controller systems installed inside an enclosure can withstand a maximum of:

a- 60°C outside the enclosure
b- 50°C outside the enclosure
c- 60°C inside the enclosure
d- 50°C inside the enclosure

10-11 If "hot spots" are generated inside the enclosure, a ______________ should be installed to help dissipate the heat.

10-12 A thermostatically controlled heater may be used in a panel if ______________ is anticipated.

10-13 True/False. A PC can operate trouble-free near an arc welding machine if it is installed in an enclosure.

10-14 Most controllers should be mounted in a ______________ position to allow maximum ______________.

10-15 The ______________ dissipates more heat than any other system component.

10-16 Which is a good location for the CPU inside the enclosure:

a- Comfortable working level
b- Adjacent to the power supply
c- Below the power supply
d- All of the above

10-17 Input/output racks are not typically placed:

a- Adjacent to the CPU
b- Beside the power supply
c- Directly above the CPU
d- In a remote enclosure

10-18 Magnetic elements are placed away from the controller components so that:

a- Power is independent
b- Space is maximized
c- Effects of noise are minimized
d- None of the above

10-19 Name at least three magnetic components that are considered during the layout design of the enclosure.

10-20 True/False. Fans or blowers should be placed at the top of the panel enclosure.

10-21 Sketch or indicate the area in which each of the following components should be mounted in the enclosure diagram shown in Figure 10-1.

1- CPU rack 64 I/O
2- I/O rack including 16 analog I/O 38 digital I/O
3- DC input/output field wiring wireway
4- AC input/output field wiring wireway
5- Power supply
6- Disconnect (30 amp)
7- Isolation transformer
8- Two MCRs
9- Fuse block
10- AC/DC wiring
11- Incoming power and field wiring
12- Disconnect
13- AC input/output signal wireway

10-22 One good reason for grouping common I/O modules is to minimize ____________________ interference.

10-23 What does the duct and wiring layout define?

10-24 True/False. The enclosure's duct and wiring layout is dependent on the placement of I/O modules within each I/O rack.

10-25 True/False. The placement of I/O modules within each I/O rack is determined during the design stage.

10-26 Incoming AC power lines should be kept ____________________ from low-level DC lines.

10-27 True/False. TTL and analog signals are considered low-level DC signals.

10-28 If the I/O wiring must cross the AC power lines, it should do so at ____________________.

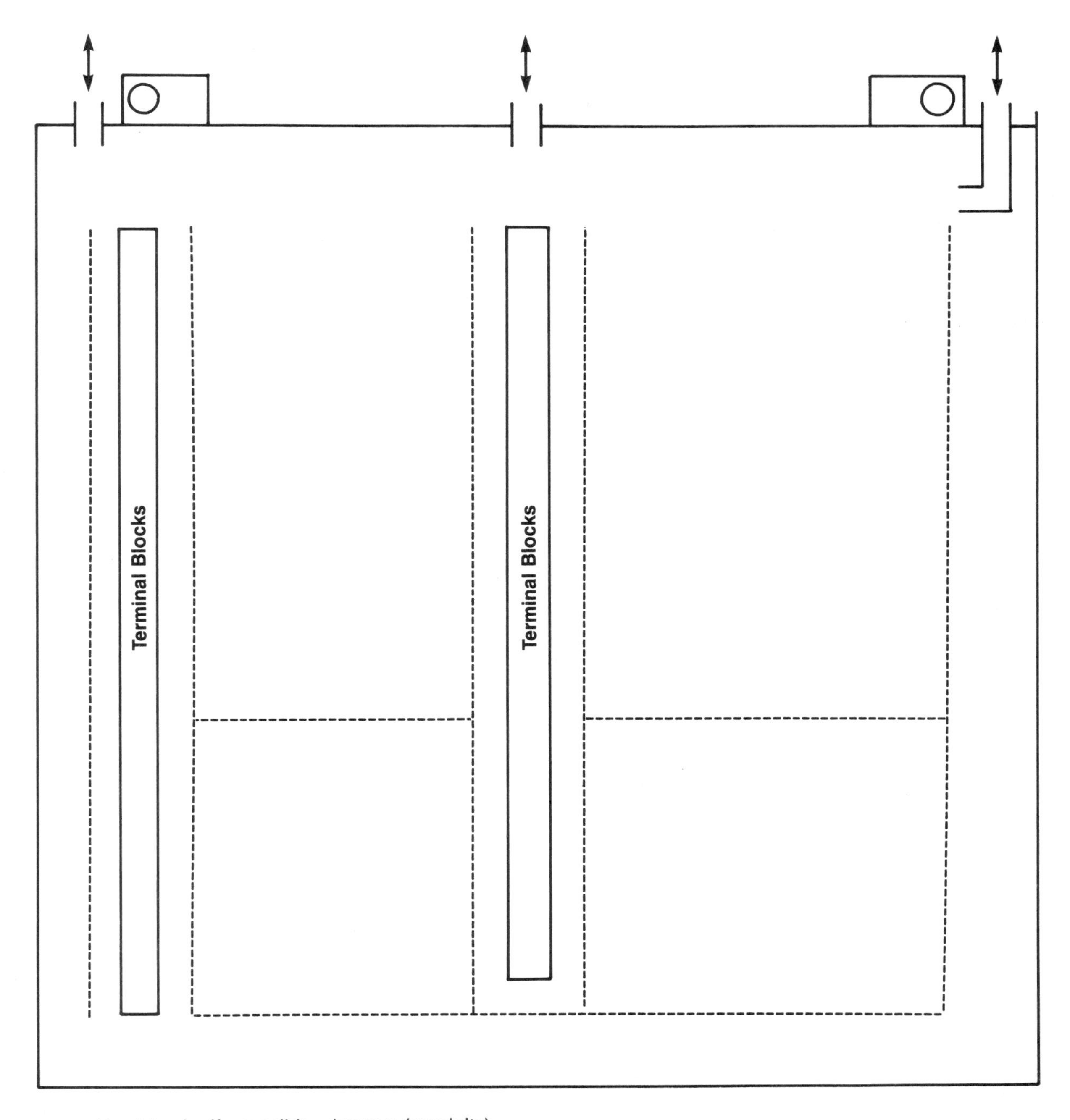

----- Used to signify possible wireways (conduits).

Figure 10-1. Problem 10-21.

10-29 The national electric code (NEC) article 250 provides data such as size, type of conductors, colors, and connections necessary for safe ________________ of electrical components.

10-30 Proper grounding procedures specify that the ground termination must be a ________________ connection.

10-31 True/False. All electrical racks should be grounded to a central ground bus.

10-32 What precaution should be taken when grounding a chassis or rack to the enclosure?

10-33 True/False. To use a common AC source for the system power supply and I/O devices is a good practice.

10-34 When is it required to use isolation transformers?

10-35 Name three devices that may produce electrical noise.

10-36 True/False. When using an isolation transformer, it is required that it provide sufficient power to the loads.

10-37 To avoid uncontrollable conditions, emergency stop switches ________________ be wired to the programmable controller.

10-38 True/False. To minimize wiring, a system should have as few emergency stops as possible.

10-39 ________________ can be used as a convenient means of removing power to the I/O system.

a- Electromechanical MCRs
b- Software MCRs
c- Software routines
d- None of the above

10-40 Briefly describe outrush, what causes it to occur, and how it can be avoided.

10-41 Temperature specifications for PCs consider typical conditions to exist when:

a- 60% of inputs are ON at one time
b- 60% of outputs are ON at one time
c- 60% of inputs and 30% of outputs are ON at one time
d- 40% of inputs and 60% of outputs are ON at one time

10-42 A ________________ transformer can be used in an installation that is subject to ________________ AC lines.

10-43 The I/O placement and wiring documents should be updated:

a- During maintenance
b- Every time there is a change
c- At the end of the project
d- During the documentation

10-44 Which of the following would generally not be considered for terminal and wire labeling?

a- Color coding
b- Using wire numbers
c- Using address numbers
d- None of the above

10-45 True/False. All wires can be bundled together as long as the bundles are kept neat.

10-46 When placing I/O modules in the I/O racks, the following should be checked:

a- Type of module
b- Slot address
c- I/O address assignment
d- All of the above

10-47 True/False. If two or more modules share the same power source, the power wiring can be jumped from one module to the next.

10-48 True/False. After terminating a wire, a gentle pull is a good practice to check for good connection.

10-49 Identify three types of devices that may require special wiring considerations during installation.

10-50 Some input devices may have a small leakage current when they are in the ____________ state.

10-51 True/False. Transistors exhibit more current leakage than triacs.

10-52 A leakage problem can occur when connecting an output module to an input module of another PC; this problem can be corrected by using a ______________________ across the input.

10-53 A MOV is used as a suppressor, and it stands for:

a- Maximum offset voltage
b- Minimum on voltage
c- Metal oxide varistor
d- None of the above

10-54 True/False. Snubber circuits are used for suppression of inductive loads.

10-55 Match the following with the appropriate drawing:

____ Small AC load suppression
____ DC load suppression
____ Large AC load suppression

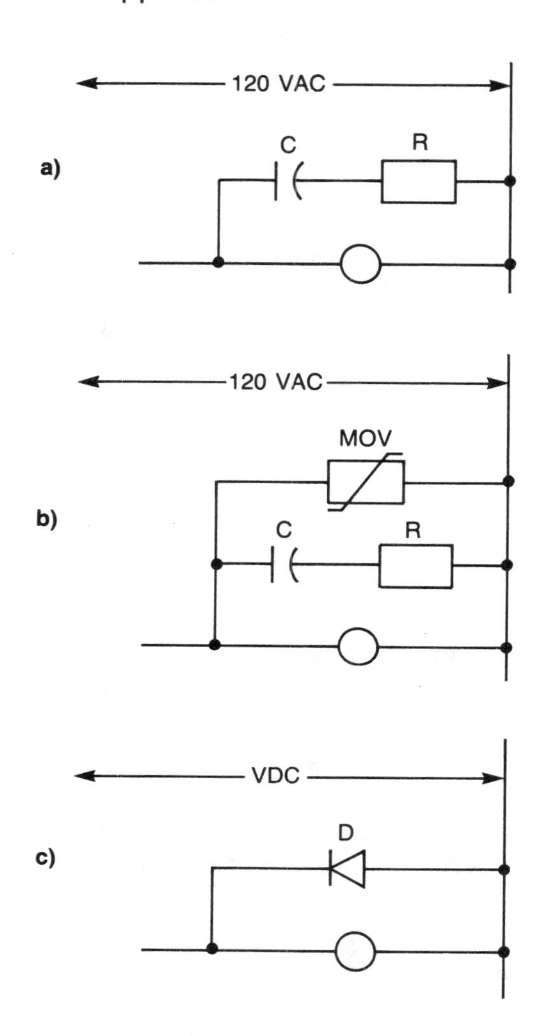

10-56 True/False. If fuses are not provided as part of the output modules, they should be installed at terminal blocks, especially when the output is driving an inductive load.

10-57 Twisted shielded cable should have:

a- At least a one-inch lay
b- Twelve twists per foot
c- Shield connected to ground at one end
d- All of the above

10-58 True/False. The static input wiring check should be performed with power applied to the controller and input devices.

10-59 How can an input device be tested so that the module's LED indicator is turned?

10-60 True/False. The static output wiring check is performed with power applied to the controller but not to the output modules.

10-61 When testing output wiring, all outputs that create mechanical motion should be ____________________.

10-62 Output devices can be tested by using the forcing function or by programming a ____________________.

10-63 Dynamic checkout assumes that:

a- Static check is completed
b- Wiring is correct
c- Software has been reviewed
d- All of the above

10-64 When should changes to the control logic be documented and stored on a permanent storage device?

10-65 When is it appropriate to perform preventive maintenance for a PC system?

10-66 When should enclosure filters be cleaned?

10-67 Why is it a good practice to clean dust build-up on heat sinks and electric circuitry?

10-68 True/False. In an environment in which vibration exists, plugs, sockets, and terminal connections should be checked periodically.

10-69 A mistake that is often repeated is leaving articles such as drawings, manuals, and other booklets on top of a CPU rack or I/O racks; this problem can cause:

a- System malfunction due to heat
b- Obstruction of air flow
c- Hot spots
d- All of the above

10-70 As a rule of thumb, the following items should be kept as spare parts for a PC system:

a- 10% of input and output modules
b- 15% of input and output modules
c- A power supply and one of each main board
d- a and c
e- b and c

10-71 If a module fuse blows repeatedly, a probable cause may be:

a- The modules output current is being exceeded
b- The output device may be shorted
c- The fuse rating may be incorrect
d- All of the above

10-72 Input modules generally have a power indicator to show that power is present at the module; however, some input modules also have a logic indicator whose function is to:

a- Show that the isolation circuit works
b- Show that logic side is ON
c- Indicate that the PC should read a logic 1
d- All of the above

10-73 What is the first check to perform when trouble-shooting an input malfunction?

10-74 An input device is connected to an input module as shown in Fig. 10-2. What should be the order of checking for an input malfunction?

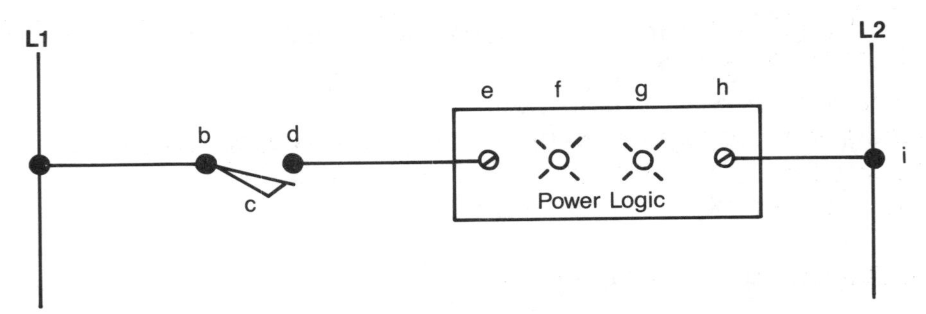

Figure 10-2. Problem 10-74.

10-75 A system has several thumbwheel switches connected to a register input module. The signal from the thumbwheel switch is in BCD. During the testing of the thumbwheel switch, it is found that the digits show up in the CRT (under monitor mode) as shown in Fig.10-3. A fault is apparent; the thumbwheel switches are changed and tested correctly; the input card is operating correctly, and there is continuity from each digit to each terminal location in the input card. What could the probable fault be?

a- Polarity is reversed in the thumbwheel switch.
b- Terminal connector at input card may be faulty.
c- Power is not being supplied to thumbwheel switch.
d- None of the above

Digit at TWS	CRT Monitor Binary	Decimal
0	0000	0
1	0001	1
2	0000	0
3	0001	1
4	0100	4
5	0101	5
6	0100	4
7	0101	5
8	1000	8
9	1001	9

Figure 10-3. Problem 10-75.

10-76 During the static check for outputs, it is necessary to test each output for continuity from the module to a terminal block, the power LED indicator, and proper connection to the field device. If the machine is not operating, reaction of the field device will not take place (a solenoid).What would be a way of finding out if the connection from the terminal block to the field device is correct?

a- Measure voltage at terminal block
b- Place a voltmeter across the load and measure the voltage
c- Observe the LED for ON, and measure the voltage
d- Turn on the machine

10-77 The key in diagnosing I/O malfunctions is to:

a- Observe the LEDs
b- Check the I/O wiring
c- Isolate the problem to the module or to the field wiring
d- Measure the voltage whether input or output

UNIT

11

DATA HIGHWAYS

- The term data highway is generally used to describe a local area network or LAN, which is defined as a high-speed medium distance communication network used to communicate between several independent PCs and/or host controllers that are located throughout a factory.

- A device connected to the highway is known as a node. Each network accommodates a different number of nodes, depending on the manufacturer.

- Industrial networks must meet the following criteria:

  - Capable of supporting real-time control
  - High data integrity
  - High noise immunity
  - High reliability in harsh environments
  - Suitable for large installations

- The main differences between business networks and industrial networks are:

  - Business networks do not require as much noise immunity.
  - The access time of business networks is less stringent.

- The most common applications of LANs are centralized data acquisition and distributed control.

TOPOLOGIES

- The topology of a network is a definition of how individual nodes are connected to the network.

- Throughput, implementation cost, and reliability are the major factors affected by topology.

- The four basic topologies presently used are: star, common bus, ring, and star-shaped ring.

- The star topology consists of one central node connected to all other nodes in the network.

- The main advantage of the star topology is that it can be easily implemented with a simple point-to-point protocol. However, a major disadvantage is that if the central node fails, the whole network communication fails.

- The common bus network is characterized by a main trunk line in which individual nodes are connected to a PC in a multi-drop fashion.

- Common topologies are very applicable to distributed systems since each station has equal independent control capabilities and can exchange information at any given time.

- In a master/slave common bus topology, communication does not take place unless it is initiated by the master, which polls the slave it wants to send/receive information from.

- The ring topology presents a major disadvantage: the failure of any node will bring the network down.
- A modification of the ring topology uses a wire center to overcome the problem of the standard ring topology; however, this arrangement requires twice as much wiring.

ACCESS METHODS

- Access method is the manner in which the PC gains access of the highway for transmitting information.
- The most commonly used access methods are polling, collision detection, and token passing.
- In the polling method, each station (slave) is polled or interrogated in sequence by a master to check if it has data to transmit.
- Collision detection, also referred to as CSMA/CD (for carrier sense multiple access with collision detection), is an access method in which each node that has a message to transmit waits until there is no traffic on the network. If two nodes try to talk at the same time, a collision occurs and retransmission will wait a variable amount of time.
- CSMA/CD does not present many problems in a data highway as long as there are not many nodes in a network.
- Token passing is an access method that is used to eliminate contention among PC stations that are trying to gain access to the highway.
- A token is a message granting a polled station the exclusive but temporary right to the highway.
- The token pass access method is preferred in applications requiring distributed control containing many nodes or having stringent response times.
- The common bus topology using the token pass access method is the most popular type of industrial network.

TRANSMISSION MEDIA

- The most common transmission media used for PC networks are twisted pair, conductors, coaxial cables, and optical fibers.
- A twisted pair conductor is a relatively inexpensive medium that has been used extensively in industry in point-to-point applications. However, it is characterized by a performance limitation over great distances due to the nonuniformity of its impedance.

- There are two main types of coaxial cables used in networks: baseband and broadband coaxial cables. Coaxial cables are extremely uniform in the cable impedance, thus making possible the achievement of greater distances and very high transmission speeds.

- Broadband coax is intended to be used in broadband networks that employ frequency division multiplexing to provide many simultaneous channels, each one having a different RF carrier frequency. Broadband coax has been used extensively for years to carry cable television signals.

- Baseband coax is used to transmit one signal at a time either modulated or unmodulated.

- Fiber optics is a medium that is totally immune to all kinds of electrical interference; however, it is very expensive, and its main shortcoming is that a low-loss tap has yet to be developed.

INTERPRETING THE SPECIFICATIONS

- The maximum number of devices that can be incorporated as a node in a network may include a PC, a vendor-supplied programmer, a host computer, or an intelligent terminal.

- There are two important lengths that must be specified in a network: the maximum length of the main line and the maximum length of each drop (between device and main line).

- When evaluating response time of a network, it is important to take the maximum response time, which includes not only the speed of the network, but also the scan times of the sending and receiving PC nodes.

- The throughput specification in LANs usually represents the number of I/O points that can be updated per second through the data highway.

- A gateway is a device that connects two or more communication networks by translating the protocols between the devices it connects.

PROTOCOLS

- A protocol is a set of rules that must be followed if two or more devices are to communicate with each other.

- Protocols define how the following problems are to be handled:

  - Communication line errors
  - Flow control to keep buffers from overflowing
  - Access by multiple devices
  - Failure detection
  - Data translation
  - Interpretation of messages

- The two most recognized protocols are the ISO (International Standards Organization) and the IEEE 802.

- The ISO's OSI (Open System Interconnection) reference model divides the various functions that protocols must perform into seven hierarchal layers. The layers are defined as follows:

  Layer 7- Application (user interface)
  Layer 6- Presentation (data conversion)
  Layer 5- Session (establishment and disconnection)
  Layer 4- Transport (end-to-end service)
  Layer 3- Network (routing)
  Layer 2- Data link (error detection, framing)
  Layer 1- Physical (electrical characteristics)

- The IEEE 802 is a protocol standard that was established by the IEEE computer society and is intended to allow communication of different manufacturers' networks. The IEEE 802 and the ISOs OSI model follow very closely the same protocol definition.

REVIEW QUESTIONS

11-1 Data highways are used to communicate between several ____________ and/or ____________ that are located throughout a factory.

11-2 A device connected to a network or highway is referred to as a ____________.

11-3 List five requirements of an industrial network.

11-4 What are the main differences between industrial networks and business networks?

11-5 What were two methods of communication between PCs before data highways came into use?

11-6 The two most common applications of data highways are ____________ and ____________.

11-7 What are three disadvantages of large applications that use a single PC system for data collection and control?

11-8 True/False. Performance and reliability are usually improved in distributed control applications when control functions are spread among several controllers.

11-9 True/False. The topology of a local area network defines how individual nodes are connected to the network.

11-10 Which of the following factors is affected by the network topology?

a- Throughput
b- Implementation cost
c- Reliability
d- All of the above

11-11 Which of the following topologies is not commonly used in PC networks?

a- Star
b- Common Bus
c- Tree
d- Ring

11-12 True/False. The only difference between the tree and star topology is the location of the network controller. (If false, describe why and state the differences.)

11-13 True/False. An advantage of the star topology is that point-to-point protocol can be easily implemented.

11-14 Which of the following factors is the greatest disadvantage of using the star topology in an industrial environment:

a- Dependance on central node speed
b- High wiring cost
c- Failure of central node will bring down the system
d- All messages must pass through the central node resulting in low throughput

11-15 Sketch the star topology showing the multiports in the main controller.

11-16 The common bus topology is characterized by a main trunkline in which individual nodes are connected to a PC in a ____________________ fashion.

11-17 Which of the following factors are advantages of the common bus topology (circle all that apply):

a- High-speed throughput
b- Each station has equal independent control
c- Reconfiguration of network is easy
d- Network controller will not bring system down

11-18 True/False. The use of a shared bus to serve all nodes is a disadvantage in the common bus topology.

11-19 True/False. A break in the trunkline in a common bus could affect at least one node, but not the network communications.

11-20 True/False. It is possible for a common bus topology to have a master/slave configuration.

11-21 How does a master controller communicate with a slave controller?

11-22 What is the major disadvantage of the ring topology?

a- Does not require multi-dropping
b- Failure of a node brings down the system.
c- It cannot use fiber-optic links
d- Collision detection cannot be avoided.

11-23 True/False. In a star-shaped ring, the use of a wire center allows failed nodes to be automatically bypassed.

11-24 The major disadvantage of a star-shaped ring network using the wire center is that:

a- A modem cannot be interfaced to the wire center.
b- A failed node will still bring the network down.
c- It requires twice as much wiring.
d- It is too complicated.

11-25 The ______________________ defines the manner in which the PC gains access to the highway.

a- Token passing
b- Access method
c- Polling
d- Introductory method

11-26 Which topology allows each of several controllers to take turns transmitting data on the network?

11-27 The access method most often used in a master-slave configuration is ______________________.

11-28 Communication using the polling access method is started by the ______________________.

11-29 The collision detection method is sometimes referred to as CSMA/CD which stands for:

a- Collision sense maximum access with collision detection
b- Collision sense multiple access with carrier detection
c- Carrier sense multiple access with collision detection
d- Carrier sense maximum access with collision detection

11-30 Explain how the collision detection access method functions and what happens when a collision occurs.

11-31 The CSMA/CD method works well as long as:

a- The distance between nodes is not too long
b- The data transmission is fast
c- There are not many nodes in the network
d- None of the above

11-32 In the CSMA/CD method, if there are many nodes in constant communication:

a- Throughput increases and access time decreases
b- Throughput drops off and access time increases
c- Throughput drops off and access time decreases
d- Throughput increases and access time increases

11-33 True/False. Collision detection is one of the most popular access methods for PC networks.

11-34 Token passing is an access technique used to eliminate:

a- Master/slave communication problems
b- Collision problems in a CSMA/CD method
c- Contention among stations that are trying to gain access to the highway
d- The station that is trying to communicate

11-35 A ______________________ is a message granting a polled station the exclusive but temporary right to control the highway.

11-36 In the token passing method, after a station (or node) finishes transmitting a message, it must ______________________ the highway right to the ______________________ node.

a- Hold
b- Keep
c- Next available
d- Next designated
e- Previously designated
f- Relinquish

11-37 True/False. The token is passed from one station to the next one in a random manner.

11-38 How are the stations identified in a common bus network that uses the token pass technique?

11-39 The token pass technique is preferred in applications requiring distributed control with:

a- Few stations
b- Many nodes
c- Many nodes and/or stringent response time
d- Fast response

11-40 True/False. A star topology can use fiber optic links as the transmission medium. If false, why?

11-41 A twisted pair is generally used as a medium for transmission of data over a network in:

a- Applications with fast baud rates
b- Point-to-point applications
c- Applications that cover long distances
d- Places with high noise

11-42 Twisted pair is limited in performance primarily due to (circle all that apply):

a- Reflection in transmission media
b- Nonuniformity of the cable impedance
c- Problems with distance
d- None of the above

11-43 Which coaxial cable is used as a medium in regular cable television signal transmission?

11-44 True/False. Baseband coax cable is more expensive than broadband coax cable.

11-45 True/False. Broadband networks use division multiplexing to provide many simultaneous channels.

11-46 Fiber optic characteristics include (circle all that apply):

a- Large bus topology usage
b- More cost
c- Small and light weight
d- Totally immune to electrical interference

11-47 True/False. The type of devices that can be supported by the network may include intelligent devices other than PCs.

11-48 What are the two lengths that must be specified in data highways?

11-49 True/False. The type of coaxial cable, i.e., RG-59, used in data highways depends on the required distance.

11-50 An important parameter to consider when evaluating the speed of a local area network would be:

a- The minimum response time
b- The maximum response time
c- The average response time
d- The throughput of the network

11-51 A ______________________ interface may be required to allow communication between two different networks or between a network and a foreign device.

11-52 True/False. Programmable controllers that support data highways generally have an interface unit that is installed in each CPU.

11-53 Information exchange on a network would not include:

a- Information stored in the executive
b- I/O point status
c- Register contents
d- PC status

11-54 What is a protocol?

11-55 What kind of information is generally defined in a protocol?

11-56 The ISO/OSI is the International Standards Organization's Open System Interconnection reference model that specifies various functions that must be performed in a protocol. Match each of the following layers with its function as specified by the model:

____ Layer 7. Application	a- Data conversion
____ Layer 6. Presentation	b- Error detection-framing
____ Layer 5. Session	c- End-to-end service
____ Layer 4. Transport	d- Routing
____ Layer 3. Network	e- User interface
____ Layer 2. Data Link	f- Electrical characteristics
____ Layer 1. Physical	g- Establishment and disconnect

11-57 True/False. The IEEE 802 protocol closely follows the ISO/OSI model.

11-58 PC manufacturers for the most part are biased toward use of the ______________________ topology using the ______________________ access method.

a- Ring
b- Master-slave
c- Common bus
d- CSMA/CD
e- Polling
f- Token pass

11-59 Match the following terms with the most appropriate answer:

____ Broadband	a- A device in the network
____ Baseband	b- Interrogation
____ Local area networks	c- Inmune to noise
____ Polling	d- Unmatch line impedance
____ Twisted pair cable	e- Interface for network
____ ISO/OSI	f- Data directly on media
____ Coaxial cable	g- Floating master
____ Gateway	h- Seven-layer model
____ Node	i- Data highway
____ Token pass	j- Data modulated on media
____ Fiber optics	k- Uniform impedance

UNIT

--]12[--

INSIGHTS TO APPLICATIONS

- The following are examples of PC applications in different areas of industry.

RUBBER AND PLASTIC

Tire Curing Press Monitoring. PCs perform individual press monitoring for time, pressure, and temperature during each press cycle. Information concerning machine status is stored in tables for later use and alerts the operator of any press malfunctions. Report generation printout for each shift includes a summary of good cures and press downtime due to malfunctions.

Tire Manufacturing. Programmable controllers can be used for tire press/cure systems to control the sequencing of events that must occur to transfer the raw tire into a tire fit for the road. This control includes molding the tread pattern and curing the rubber to obtain the road resistant characteristics. This PC application substantially reduces the space required and increases reliability of the system and quality of the product.

Rubber Production. Dedicated programmable controllers provide accurate scale control, mixer logic functions, and multiple formula operation of carbon black, oil, and pigment used in the production of rubber. The system maximizes utilization of machine tools during production schedules, tracks in-process inventories, and reduces time and personnel required to supervise the production activity and the manually produced shift-end reports.

Plastic Injection Molding. A PC system controls variables, such as temperature and pressure, that are used to optimize the injection molding process. The system provides closed-loop injection such that several velocity levels can be programmed to maintain consistent filling, reduce surface defects and stresses, and shorten cycle time. The system can also accumulate production data for future use.

CHEMICAL AND PETROCHEMICAL

Ammonia and Ethylene Processing. Programmable controllers monitor and control large compressors that are used during the manufacturing of ammonia, ethylene, and other chemicals. The PC monitors bearing temperatures, operation of clearance pockets, compressor speed, power consumption, vibration, discharge temperatures, pressure, suction flow, and gas composition.

Dyes. PCs monitor and control the dye processing used in the textile industry. They provide accurate processing of color blending and matching to predetermined values.

Chemical Blending. The PC controls the batching ratio of two or more materials in a continuous process. The system determines the rate of discharge of each material, in addition to providing inventory records on other useful data. Several batch recipes can be logged and retrieved automatically or on command from the operator.

Fan Control. PCs automatically control fans based on levels of toxic gases in a chemical production environment. This system provides effective measures of exhausting gases when a level of contamination is reached. The PC controls the fan start/stop, cycling, and speeds so that safety levels are maintained while energy consumption is minimized.

Gas Transmission and Distribution. Programmable controllers monitor and regulate pressures and flows of gas transmission and distribution systems. Data is gathered and measured in the field and transmitted to the PC system.

Oil Fields. PCs provide on-site gathering and processing of data pertinent to characteristics such as the depth and density of drilling rigs. The PC controls and monitors the total rig operation and alerts the operator of any possible malfunction.

Pipeline Pump Station Control. PCs control mainline and booster pumps for the distribution of crude oil. Measurement of flow, suction, discharge, and tank low and high limits are some of the functions they fulfill. Possible communications with SCADA (Supervisory Control And Data Acquisition) systems can provide total supervision of pipeline.

POWER

Plant Power System. The programmable controller regulates the proper distribution of available electricity, gas, or steam. In addition, the PC monitors power house facilities, schedules distribution of energy, and generates distribution reports. The PC controls the loads during operation of the plant, as well as the automatic load shedding or restoring during power outages.

Energy Management. Through the reading of inside and outside temperatures, the PC controls heating and cooling units in a manufacturing plant. The PC system controls the loads, cycling them during predetermined cycles and keeping track of how long each load should be on during the cycle time. The system provides scheduled reports on the amount of energy used by the heating and cooling units.

Coal Fluidization Process. The controller can monitor how much energy is generated from a given amount of coal and regulate the coal crushing and mixing with crushed limestone. The PC monitors and controls burning rates, temperatures generated, sequencing of valves, and analog control of jet valves.

Compressor Efficiency Control. PCs are used to control several compressors located at a typical compressor station. The system handles safety interlocks, start-up/shutdown sequences, and compressor cycling and keeps the compressors running at maximum efficiency using the non-linear curves of the compressors.

METALS

Steelmaking. The PC controls and operates furnaces and produces the metal in accordance with preset specifications. The controller also calculates oxygen requirements, alloy additions, and power requirements.

Loading and Unloading of Alloys. Through accurate weighing and loading sequences, the system controls and monitors the quantity of coal, iron ore, and limestone to be melted. The unloading sequence of the steel to a torpedo car can also be controlled.

Continuous Casting. PCs direct the molten steel transport ladle to the continuous-casting machine, where the steel is poured into a water-cooled mold for solidification.

Cold Rolling. PCs are used to control the conversion of semi-finished products into finished goods through cold rolling machines. The system controls the speed of the motors to obtain correct tension and provide adequate gauging of the rolling material.

Aluminum Making. Controllers monitor the refining process in which impurities are removed from bauxite by heat and chemicals. The system can grind and mix the ore with chemicals and then pump them into pressure containers, where they are heated, filtered, and combined with more chemicals.

PULP AND PAPER

Pulp Batch Blending. The PC controls sequence operation, quantity measurement of ingredients, and storage of recipes for the blending process. The system allows operators to modify batch entries of each quantity if necessary and provides hard copy printouts for inventory control and for accounting of ingredients used.

Batch Preparation for Paper Making Process. Applications include control of the complete stock preparation system for paper manufacturing. Recipes for each batch tank are selected and adjusted via operator entries. The system can also control the feedback logic for chemical addition based on tank level measurement signals. At the completion of each shift, the PC system provides management reports for material usage.

Paper Mill Digester. PC systems provide complete control of pulp digesters for the process of making pulp from wood chips. The system calculates and controls the amount of chips based on density and the digester volume; the percent of cooking liquors is calculated, and the required amounts are added in sequence. The PC ramps and holds the cooking temperature until the cook is completed. All data concerning the process is then transmitted to the PC for reporting.

Paper Mill Production. The controller regulates the average basis weight and moisture variable for paper grade. The system manipulates the steam flow valves, adjusts the stock valves to regulate weight, and monitors and controls total flow.

GLASS PROCESSING

Annealing Lehr Control. PCs control the Lehr machine used to remove the internal stress from glass products. The system controls the operation by following the annealing temperature curve during the reheating, annealing, straining, and rapid cooling processes through different heat and cool zones. Improvements are made in the ratio of good glass to scrap, reduction in labor cost, and energy utilization.

Glass Batching. PCs control the batch weighing system according to stored glass formulas. The system also controls the electromagnetic feeders for infeed to and outfeed from the weigh hoppers, manual shut-off gates, and other equipment.

Cullet Weighing. PCs direct the cullet system by controlling the vibratory cullet feeder, weight-belt scale, and shuttle conveyor. All sequences of operation and inventory of quantities weighed are kept by the PC for future use.

Batch Transport. PCs control the batch transport system including reversible belt conveyors, transfer conveyors to cullet house, holding hoppers, shuttle conveyors, and magnetic separators. The controller takes action after discharge from the mixer and transfers the mixed batch to the furnace shuttle, where it is discharged to the full length of the furnace feed hopper.

MATERIALS HANDLING

Storage and Retrieval Systems. A PC is used to load parts and carry them in totes in the storage and retrieval system. The controller keeps tracking information such as storage lane number where parts are stored, the parts assigned to specific lanes, and quantity of parts in any particular lane. This PC arrangement allows rapid changes of parts loaded or unloaded from the system. The controller also provides inventory printouts and informs the operator of any malfunctions.

Automatic Plating Line. The PC controls a set pattern for the automated hoist, which can traverse left, right, up, and down through the various plating solutions. The system knows at all times where the hoist is located.

Conveyor Systems. The PC controls all the sequential operations, alarms, and safety logic necessary to load and circulate parts on a main line conveyor as well as sorting product to the correct lane. The PC can also schedule lane sorting to optimize palletizer duty. Records are kept on a shift basis for production of good parts and part rejections if required.

Automated Warehousing. The PC controls and optimizes the movement of stacking cranes and provides high turn-around of materials requests in an automated high-cube vertical warehouse. The PC also controls aisle conveyors and case palletizers to significantly reduce manpower requirements. Inventory control figures are maintained and can be provided upon request.

AUTOMOTIVE

Internal Combustion Engine Monitoring. The system acquires data recorded from sensors located at the internal combustion engine. Measurements taken include water temperature, oil temperature, RPMs, torque, exhaust temperature, oil pressure, manifold pressure, and timing.

Carburetor Production Testing. PCs provide on-line analysis of automotive carburetors in a production assembly line. The systems significantly reduce the test time, while providing greater yield and better quality carburetors. Some of the variables that are tested are pressure, vacuum, and flow.

Monitoring Automotive Production Machines. The system monitors total parts, rejected parts, parts produced, machine cycle time, and machine efficiency. All statistical data are available to the operator and also at the end of each shift.

Power Steering Valve Assembly and Test. The PC system controls a machine to insure proper balance of the valves and maximize left and right turning ratios.

MANUFACTURING/MACHINING

Production Machines. The PC controls and monitors automatic production machines at high efficiency rates. The piece-count production and machine status are also monitored. Corrective action can be taken immediately if a failure is detected by the PC.

Transfer Line Machines. PCs monitor and control all transfer line machining station operations and the interlocking between each station. The system receives input from the operator to check the operating conditions of line-mounted controls and reports any malfunctions. This arrangement provides greater machine efficiency, higher quality products, and lower scrap levels.

Wire Machine. The controller monitors the time and ON/OFF cycles of a wire drawing machine. The system provides ramping control and synchronization of electric motor drives. All cycles are recorded and reported on demand to obtain the machine's efficiency as calculated by the PC.

Tool Changing. The PC controls a synchronous metal-cutting machine with several tool groups. The system keeps track of when each tool should be replaced, based on the number of parts it manufactures. It also displays the count and replacements of all the tool groups.

Paint Spraying. PCs control the painting sequences in auto manufacturing. Style and color information is entered by the operator or a host computer and is tracked through the conveyor until the part reaches the spray booth. The controller decodes the part information and then controls the spray guns to paint the automotive part. The spray gun movement is optimized to conserve paint and to increase part throughput.

REVIEW QUESTIONS

12-1 List five applications of the glass industry in which programmable controllers are used.

12-2 List three applications in the chemical processing industry in which PCs are used.

12-3 How can a programmable controller be used in a conveyor system that transports materials from production to the finished goods warehouse?

12-4 Name some application in the automotive industry in which PCs may be used.

12-5 Emergency diesel generators are usually employed in situations that require generation of electrical power after a power down. These generators typically go through a sequential routine of events. Describe how you might apply a programmable controller with a DC power supply to solve the sequential start-up of an emergency generator.

12-6 Why are programmable controllers used in the modernization of an existing machine or production line?

12-7 Why would a PC system be useful in a batching application?

12-8 What are some of the benefits that a machine manufacturer (OEM) would acquire from the use of programmable controllers?

12-9 Explain why some applications are best applied using distributed control as opposed to centralized control.

12-10 Explain why a small application could be justified if the PC is only used to replace timers.

UNIT --]13[--

SELECTING THE RIGHT PROGRAMMABLE CONTROLLER

PRODUCT RANGES

- Programmable controllers are divided into four major areas: small, medium, large, and very large. These areas of segmentation are based primarily on the I/O count capacity of the system and its complexity.

- Within the four areas, three overlapping sections include products of one section with features of products of the next range category.

- As the I/O count increases, so does the complexity and cost of the system. The complexity implies more sophisticated instructions, larger memory, and a greater variety of I/O modules available.

- The first segment, small PCs, is characterized by systems having less than 128 I/O and generally applied in ON/OFF type of control. Microcontrollers (less than 32 I/O) are included in this segment.

- The second segment, medium PCs, is composed of PCs that range between 128 I/O and 1024 I/O and include analog control, data manipulation, and arithmetic capabilities.

- The third segment, large PCs, include systems in the range of 512 I/O to 4096 I/O and are used in applications requiring extensive data manipulation, acquisition, and reporting. Products in this segment have local area networks and higher arithmetic operations.

- The fourth segment, very large PCs, includes systems that have capabilities to handle from 2048 I/O to 8192 I/O. These larger systems are used if a large application requires centralized controls or if a central unit is used as a supervisory PC. This segment is characterized by PCs having large memory systems and host computer communication capability.

- The overlapping sections are located between the first and second, second and third, and third and fourth segments. The area of overlapping is marked by the ranges of the I/O count between any two of the major segments.

DEFINING THE CONTROL SYSTEM

- During the definition of the control system, the following items must be considered and evaluated:

  - Input/output
  - Type of control
  - Memory
  - Software
  - Peripherals
  - Physical and environmental

- Determination of the amount of I/O required is typically the first problem addressed during the selection of a controller.
- The type of discrete I/O, analog I/O, or special I/O required has to be investigated so that the input/output system is appropriate for the application.
- Remote I/O subsystems must be considered if the field devices are located remotely from the processor.
- Three control schemes are used in a programmable controller system: individual (or segregated) control, centralized control, and distributed control.
- Individual control is used when a PC controls a single machine.
- Centralized control is used when several machines or processes are controlled with one central PC.
- Distributed control involves two or more PCs controlling several stations and communicating information through a local area network.
- The two most important considerations concerning the system memory are the type and amount of memory required.
- Software considerations that must take place include the type of instructions needed to accomplish the control program. The system software will determine the degree of difficulty that would be required to implement the user software.
- Peripheral devices must be evaluated to determine what type would be required for what function. The first peripheral to be evaluated, however, is the programming device.
- Physical and environmental characteristics of various controllers must be evaluated and checked to be sure the system will comply with the environment in which it will be operating.
- Product reliability plays an important role in the overall system performance.
- The burn-in procedure that manufacturers have in their products has to be investigated. This burn-in procedure eliminates the infant mortality in their products, thus minimizing field failures.
- Plant standardization should be a consideration during the selection of the controller. If future applications will be done at the same location, standardization on one or two products will mean fewer spare parts in inventory and less retraining on equipment.

- The following steps are a summary of the major considerations involving the programmable controller selection:

  Step 1: Know the process to be controlled.
  Step 2: Determine the type of control.
  Step 3: Determine the I/O interface requirement.
  Step 4: Determine software languages and functions.
  Step 5: Consider the type of memory.
  Step 6: Consider memory capacity.
  Step 7: Evaluate processor scan time requirements.
  Step 8: Define programming and storage device requirements.
  Step 9: Define peripheral requirements.
  Step 10: Determine any physical and environmental constraints.
  Step 11: Evaluate other factors that can affect selection.

REVIEW QUESTIONS

13-1 Low-end programmable controllers are small size PCs that are designed to be used as ____________________.

13-2 Which of the following PC categories is most appropriate for supervisory control and hierarchal systems?

a- Small PCs
b- Medium PCs
c- Large PCs
d- Very large PCs
e- b and c
f- c and d

13-3 Describe briefly each category of products (i.e.,1, 2, 3, 4 and a, b, c) of the chart in Fig. 13-1.

13-4 What is the primary specification for categorizing programmable controllers?

13-5 True/False. As the I/O count increases in a system, so does the cost but not the memory capacity.

13-6 True/False. The shaded areas A, B, and C (Fig. 13-1) represent products of a given category that have feature enhancements over the standard products for that category.

13-7 Micro-PCs are generally in which I/O range?

a- 64 I/O or less
b- 128 I/O
c- 32 or more I/O
d- 32 or less I/O

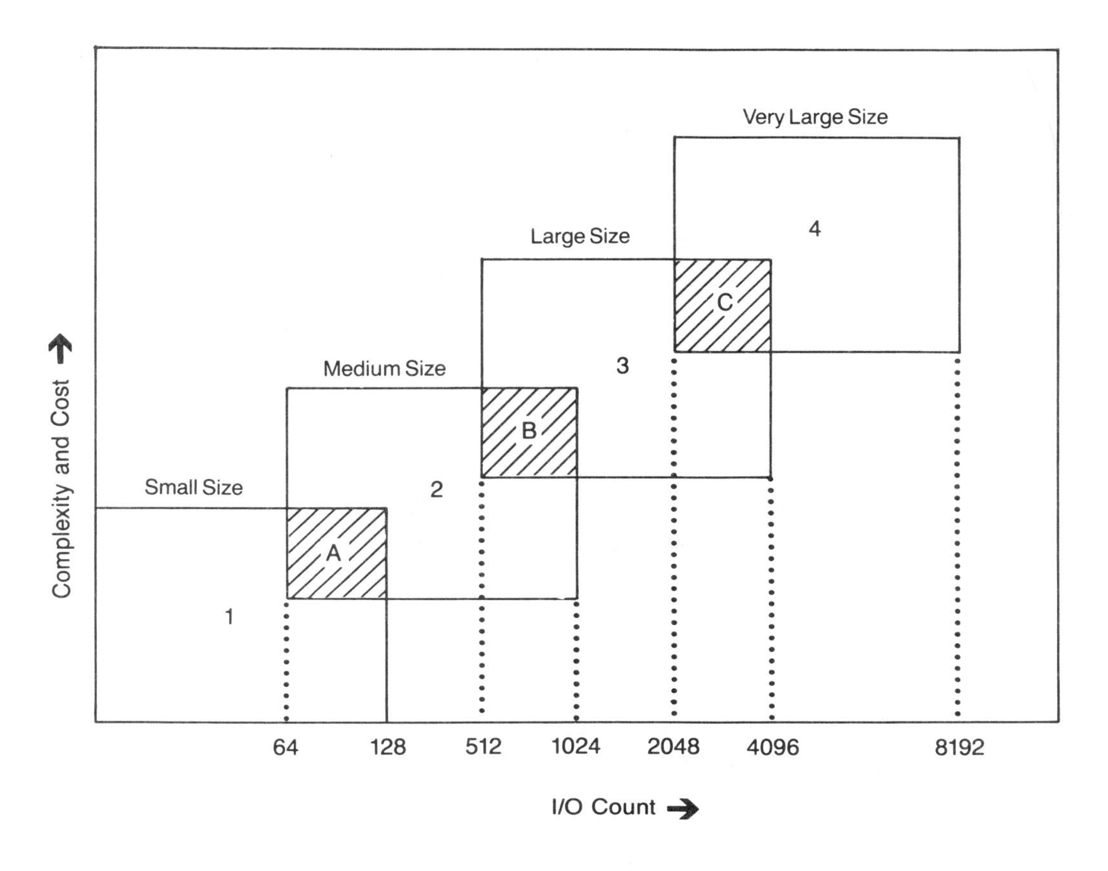

Figure 13-1. Problem 13-6.

13-8 True/False. A micro PC is likely to fall in category A of Fig. 13-1.

13-9 Products in area A of Fig.13-1 also offer features that are standard for products in area:

a- 1
b- 2
c- Some of B
d- None of the above

13-10 Products in area B include:

a- Enhancement of standard features of products in segment 1
b- Enhancement to standard features of products in segment 2
c- Features of area C
d- Enhancements of area A

13-11 Medium-size PCs are applied when:

a- True/False. More than 128 I/O is required
b- True/False. Less than 2048 I/O is required
c- True/False. Analog control is needed
d- True/False. 32K of memory is needed
e- True/False. Data manipulation is required

13-12 Name five features that are generally available in large PCs.

13-13 Very large controllers are used:

a- In distributed control applications
b- In centralized control applications
c- As a host
d- All of the above
e- None of the above

13-14 True/False. When selecting a PC, one should consider only the current application requirements.

13-15 Which of the following items need not be evaluated and defined during the selection of a PC:

a- Inputs/outputs
b- Type of control
c- Documentation
d- Peripherals

13-16 What is generally the first hardware consideration when trying to select a PC?

13-17 True/False. During the determination of I/O requirements, expansion should be considered.

13-18 Name at least three factors that must be considered when evaluating discrete outputs.

13-19 When a digital output module has fuses incorporated in the module, attention should also be placed on:

a- The current requirement
b- The accessibility
c- The indicators
d- None of the above

13-20 If an application calls for different power sources for output devices, one must consider modules with:

a- Proper ratings
b- Fuses
c- Isolated commons
d- Common returns

13-21 Match the following terms with the appropriate answer:

____ Bipolar	a- Fast input, positioning
____ Special analog input	b- Current output
____ 4 to 20 mA	c- −10 to +10 VAC
____ Special I/O	d- RTD
____ PID module	e- Measurement of flow, temperature
____ Analog I/O	f- Processing in module

13-22 Remote inputs and outputs should be considered to ____________________ wiring system when the distance between I/O and field devices is very long.

13-23 Sketch each of the following control system configurations: a)individual, b)centralized, and c)distributed.

13-24 True/False. Individual control is the same as segregated control.

13-25 Which of the following statements does not apply to individual control:

a- Used to control a single machine
b- Used to control more than one machine
c- Could have a few remote I/O
d- Can be applied to an injection molding machine

13-26 True/False. Centralized control is used when several machines or processes are controlled by one programmable controller.

13-27 An advantage of centralized control is that:

a- It has short control programs
b- It avoids problems of decentralizing the control task
c- It minimizes wiring
d- It has expansion capability

13-28 A ____________________ system can be used in centralized control applications that require a "hot" back-up.

13-29 A distributed control configuration involves at least:

a- A large PC
b- One mile communication distance
c- Two PCs that communicate with each other
d- Special I/O modules

13-30 The best method of communication in a distributed control system is via a ____________________.

13-31 True/False. Communication between two different data highways is easily implemented.

13-32 The two most important considerations when evaluating the system memory are the ________________ and ________________ of memory.

13-33 True/False. In general, small PCs have a fixed amount of memory.

13-34 The memory requirements are a function of ________________, and ________________.

- a- Inputs
- b- Outputs
- c- Inputs and outputs
- d- Complexity of the control program

13-35 True/False. The instruction set selected in a programmable controller will affect the degree of difficulty in implementing the software program.

13-36 What is the first peripheral that must be considered?

13-37 True/False. Peripheral requirements should be evaluated along with the CPU. If true, why?

13-38 True/False. The physical and environmental characteristics of a system have little impact on the total system reliability.

13-39 Which of the following should be taken into consideration when specifying the controller and I/O system? (circle all that apply)

- a- Size of the controller
- b- Operating parameters
- c- Packaging
- d- All of the above

13-40 True/False. Technical support from a manufacturer is not important if the engineering group is very capable.

13-41 Technical support should begin:

- a- After the purchase of equipment
- b- Before the purchase of equipment
- c- After a good lunch
- d- After commitment to purchase is made

13-42 True/False. The controller's reliability plays an important role in the overall performance of the control system.

13-43 Describe a typical burn-in procedure for PCs and its purpose?

13-44 ______________________ of a product line should be considered when possible as part of the PC selection decision.

a- Looks
b- The manufacturing process
c- Standardization
d- The salesman

13-45 Programmable controller families generally share:

a- The same I/O structure
b- Programming device
c- Elementary instruction set
d- All of the above

13-46 Arrange the following considerations (a-k) for selecting a PC in the order in which they would likely be considered (step 1 thru step 11).

____ Step 1	a- Know the process to be controlled
____ Step 2	b- Memory tape
____ Step 3	c- Type of control
____ Step 4	d- Memory capacity
____ Step 5	e- I/O interface requirements
____ Step 6	f- Scan requirements
____ Step 7	g- Software language and functions
____ Step 8	h- Physical and environmental constraints
____ Step 9	i- Programming and storage devices
____ Step 10	j- Peripherals requirements
____ Step 11	k- Evaluate other factors

SECTION

--]II[--

ANSWERS TO REVIEW QUESTIONS

UNIT
--] 1 [--

ANSWERS

1-1 The design criteria for the first programmable controller was defined and specified in 1968 by the Hydramatic Division of General Motors Corporation.

1-2 d- Programmable controller. Of course, PC does stand for the other choices also in other uses.

1-3 True

1-4 b- ON/OFF control. Primarily designed to substitute hardwired relay controls.

1-5 It would eliminate the high cost associated with inflexible relay control systems and take advantage of the reliability offered by solid-state technology. With an architecture similar to a computer, it would have flexible programming and modular components. It would be capable of sustaining the industrial environment.

1-6 b- Microprocessor technology. In the early 1970s, micros added great flexibility and intelligence to PCs.

1-7 d- All of the above

1-8 True

1-9 True. New enhancements in software made easier the utilization of the hardware available including the data acquisition and manipulation.

1-10 a- Fast scan time
b- Functional blocks
b- Data handling
a- Intelligent I/O
a- High-density I/O
b- High-level languages
a- Small, low-cost PCs

1-11 The central processing unit, input/output interface

1-12

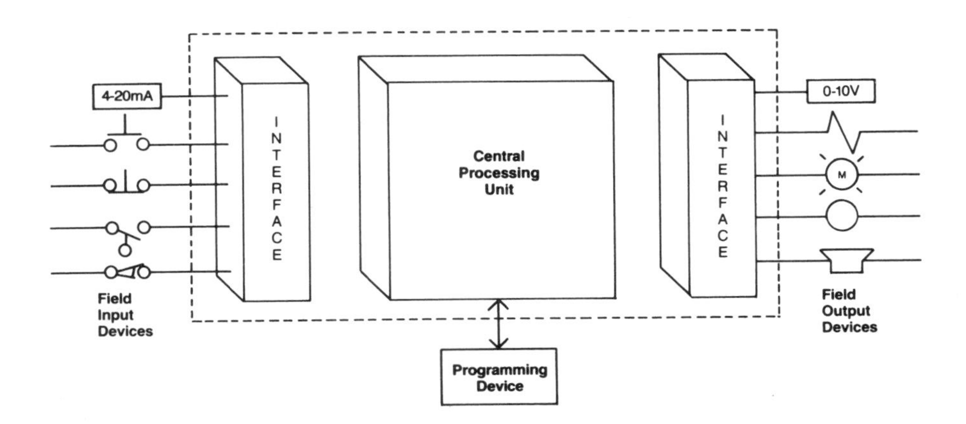

1-13

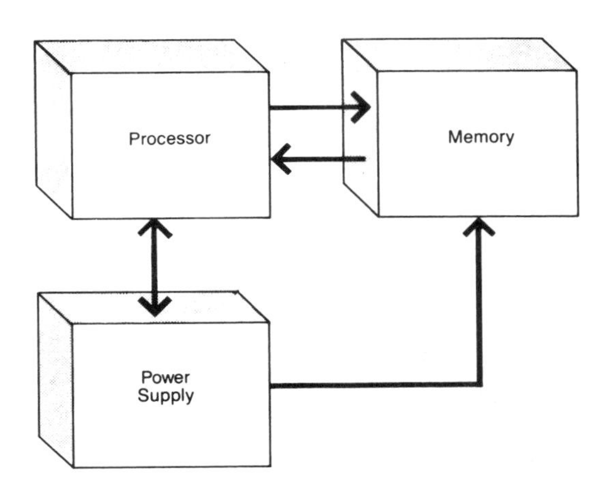

1-14 True

1-15 Condition, received, sent

1-16 Programming

1-17 CRT

1-18 Scanning

1-19 PCs are designed to survive the industrial environment, which includes high noise, electro-magnetic interference, vibration, humidity, and high temperatures (60°C). The PC is also designed to be software-friendly not only to engineers and designers, but also to technical electricians.

1-20 Microcontrollers

1-21 True. The I/O count is the first parameter generally looked at and, therefore, the primary basis for product segmentation.

1-22 True

1-23 Programmable. A PC can be reprogrammed for a new application very easily since all the control program is stored in a programmable memory.

1-24 False. The only connection between the input and output devices is through the control program.

1-25 False. PCs do not take half of the space of a small, hardwired relay control panel. The more complex the system, the less, proportionally, the space required.

1-26 The use of remote input and output subsystems will greatly reduce the amount of wiring required. This is true because communication with a subsystem is simply done through two pairs of wires as opposed to bringing all wires to the main PC. Start-up is also simplified since any wiring problems could be tracked easily.

1-27 True. The PC can be programmed to annunciate certain failures such as the malfunction of a solenoid valve.

1-28 b- Minimal space requirements
e- Expandability
a- High reliability
c- Easily changed presets
d- Multi-function capabilities
f- Reduced trouble-shooting

1-29 True

1-30 True. This is one of the reasons PCs are used.

UNIT
--]2[--

ANSWERS

2-1 True

2-2 e- None of the above. All number systems have a base, are used for counting, represent quantities, and have a set of symbols.

2-3 True

2-4 The base minus one. For instance, a number system with a base of 13 will have 13 symbols where the decimal equivalent of the largest-valued symbol is 12. These symbols can be numerical, alphabetical, or a combination of both.

2-5 c- Hexadecimal
d- Binary
b- Octal
a- Decimal

2-6 d- 10
f- 2
e- 15
a- 9
b- 7
c- 1

2-7 True. The position of each digit or symbol in a number system has a weight that is equal to its base to the power of its position (starting from right to left with position zero).

2-8 The decimal number system

2-9 b- 14 in decimal. Fourteen is a given name to decimal 14. In octal, 14 is a one-four octal.

2-10 The decimal equivalent will be:

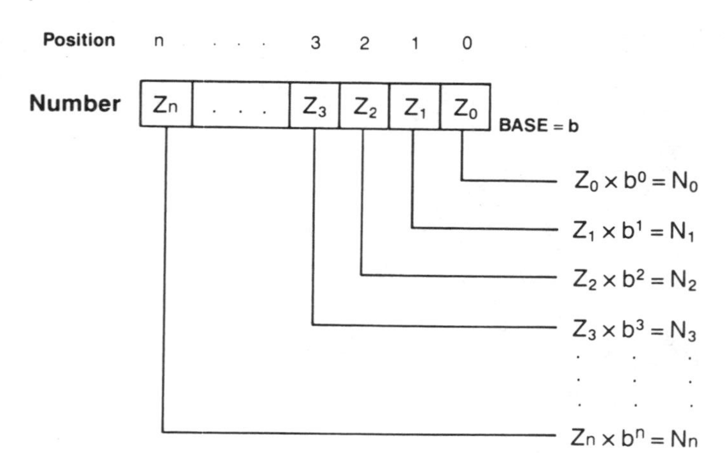

The sum of N_0 through N_0 will the decimal equivalent of the number in base "b."

2-11 The binary system was adopted for convenience, since it is easier to design digital systems that can distinguish two states as possible values rather than ten as decimal. Most discrete devices have two states, ON (1) or OFF (0).

2-12 a- 155
b- 101
c- 219
d- 85

2-13 b-Zero and carry one.

```
  1  = decimal 1
 +1  = decimal 1
 ----
 10  = decimal 2
 └────carry
```

2-14 A bit is one digit in the binary number system. Bit stands for BInary digiT.

2-15 In general, a nibble is a group of four bits, a byte is a group of eight bits, and a word is a group of one or more bytes. However, a 12 bit word is a valid word length.

2-16

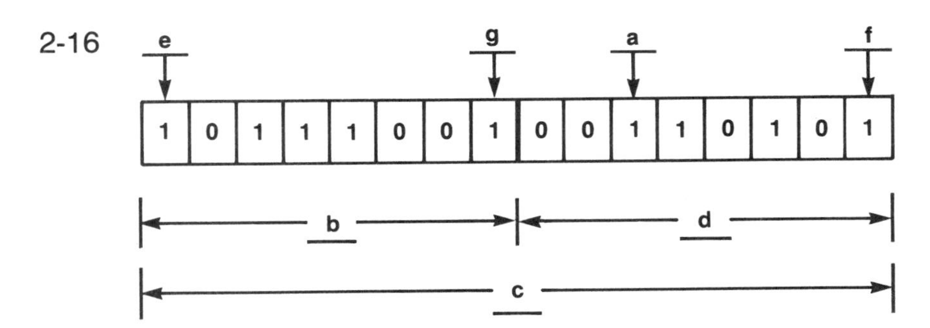

2-17 True

2-18 False. In any number system you can count to any decimal equivalent. All number systems are used for counting.

2-19 377 octal; in binary, it will be 11111111.

2-20 d- 0101
b- 0111
e- 8 decimal
a- 12 octal
c- 17 octal

2-21 177777 octal; in binary, it will be 1 111 111 111 111 111.

2-22 10 hexadecimal

2-23 2B9 will require at least ten bits to represent it as 10 1011 1001.

2-24 The maximum hexadecimal that can be represented in a two byte word (16 bits) is FFFF, binary 1111 1111 1111 1111.

2-25 Since 27 = 3^3 we can convert $IG4_{27}$ by writing the 3-bit base 3 equivalent of each digit in base 27. Therefore, $IG4_{27}$ will be 200121011_3. Another method (long way) will be to change $IG4_{27}$ to decimal and then convert from decimal to base 3:

Base 27 to decimal:

I G 4_{27}

4×27^0	=	$4_{10} \times 27^0$	=	4
$G \times 27^1$	=	$16_{10} \times 27^1$	=	432
$I \times 27^2$	=	$18_{10} \times 27^2$	=	13122
				13558_{10}

Decimal to base 3:

3 | 13558	Remainders
3 | 4519	200121011_3
3 | 1506	
3 | 502	
3 | 167	
3 | 55	
3 | 18	
3 | 6	
3 | 0	

2-26 Using the same procedure as in the previous problem, $A3_{16}$ will be 163_{10} and 2203_4.

2-27 AA_{16} must first be converted to decimal and then to base 3 to obtain the 2022_3 result.

Base 16 to decimal:

A A_{16}

$A \times 16^0$	=	$10_{10} \times 16^0$	=	10
$A \times 16^1$	=	$10_{10} \times 16^1$	=	160
				170_{10}

Decimal to base 3:

3 | 170	Remainders
3 | 56	2022_3
3 | 18	
3 | 6	
3 | 0	

2-28 a- 153 octal can be converted directly to binary by using the three-bit binary equivalent of octal; therefore, the solution will be $001\ 101\ 011_2$.

b- F35 hexadecimal can be converted to its binary equivalent by using the four-digit equivalent of hexadecimal and then to octal by grouping the three-bit octal equivalents. The solution, therefore, will be 111100110101_2 which is 7465_8.

c- 11100_2

Decimal to binary:

2 | 28	Remainders
2 | 14	11100_2
2 | 7	
2 | 3	
2 | 1	
2 | 0	

d- 43_8

Decimal to octal:

8 | 35	Remainders
8 | 4	43_8
8 | 0	

e- 45_{10}

Binary to decimal:

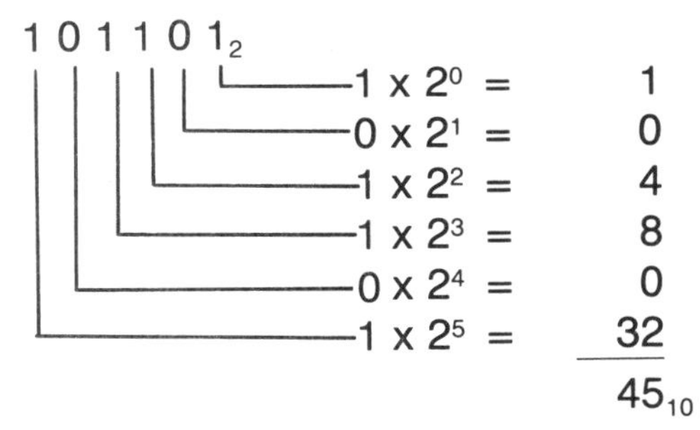

f- 38_{10}

Octal to decimal:

46

$6 \times 8^0 = 6$

$4 \times 8^1 = 32$

38_{10}

2-28 Continued

g- $2F_{16}$

Octal to hexadecimal:

$57_8 = 101\ 111_2 = 0010\ 1111$
2 F_{16}

2-29 True. To complement a number is to obtain its negative.

2-30 Addition. The only arithmetic operation that digital devices can perform is addition. Subtraction is an addition once the complement is obtained; multiplication and division are performed through successive additions.

2-31 True. The number 20 in binary requires five bits (10100); if we are using signed arithmetic, an extra bit would be used to represent the sign. Therefore -20 would require six digits or binary bits.

2-32 b- Maximum of +177 octal; minimum of 10000001 binary. A byte has eight bits; in two's complement arithmetic the most significant digit will be the sign bit. The maximum positive number will be 01111111_2 or 177 octal.The two's complement of 01111111_2 will be 10000001_2 which corresponds to -177 octal (in two's complement).

2-33 Maximum +32767 and minimum -32767. The most significant bit is the sign bit.

2-34 The number +32767 is $0111\ 1111\ 1111\ 1111_2$ and -32767 is $1000\ 0000\ 0000\ 0001_2$ using two's complement arithmetic.

2-35 a-Alphanumeric and control characters are both represented in the ASCII code.

2-36 American Standard Code for Information Interchange

2-37 ASCII uses 7 bits or 8 bits when parity is used. The parity bit is the most significant bit.

2-38 d- 200 octal. There are 128 possible characters in ASCII, from 000 to 177 octal (0 to 127 decimal) or a total count of 128 or 200_8 (including the 0).

2-39 Four bits are required to represent the number 7 in BCD (0111).

2-40 The binary pattern of a BCD number is not equal to the binary pattern of the number the BCD code represents in decimal. The number 8796 in BCD is expressed in binary as 1000 0111 1001 0110, which is equivalent to 34710 decimal.

2-41 Only one bit at a time

2-42 Encoders generally use the Gray code to determine angular position.

2-43 True. Registers and words are generally used interchangeably in PCs.

2-44 True. Even though a word is one or more bytes, it can also have less bits than a byte.

2-45 In one register (16 bits), there can be four BCD digits, one per every four bits, with a maximum value of 9999 BCD. The maximum BCD number a BCD digit can represent is 9 (1001).

UNIT

--]3[--

ANSWERS

3-1 c- The two-state concept conditions. This concept includes not only the ON/OFF state of lights or voltage levels, but also the state of all digital or discrete devices that exhibit this characteristic.

3-2 a- Positive logic. In digital systems, the positive logic concept is the most commonly used.

3-3 d- All of the above

3-4 b- Multiplication. The AND function of two signals can be thought of as a multiplication, since the results will be 1 if both inputs are one; 1x1 = 1, 1x0 = 0.

3-5 False. Logic gates were actually derived from the Boolean expressions.

3-6 True. The ladder diagrams in a PC are also called contact symbology because of the similar graphic representation.

3-7 False. In general, most output rungs can only control one output at a time in a programmable controller. However, very few controllers will allow multiple output rungs.

3-8 c- Inverter. The NOT will invert the input signal; if the input is a 1, the output will be 0 and viceversa.

3-9 a- The normally-open contact. When evaluated by the program, this symbol is examined for a "1" to close the contact; therefore, the signal referenced by the symbol must be ON, closed, activated, etc.

b- The normally closed contact. When evaluated by the program, this symbol is examined for a "0" to keep the contact closed; therefore, the signal referenced by the symbol must be OFF, open, de-activated, etc.

c- Outputs. An output on a given rung will be energized if any left-to-right path has all contacts closed, with the exception of power flow going in reverse before continuing to the right. An output can control a connected device if the reference address is also a termination point or internal output used exclusively within the program.

d- Repeated use of contacts. A given input,output,or internal output can be used throughout the program as many times as required.

3-10 b- Deactivate. In positive logic, a 1 will activate a device and a 0 will deactivate the device. If they are passed through a NOT condition, a 1 will deactivate a device since the NOT output will be a 0.

3-11

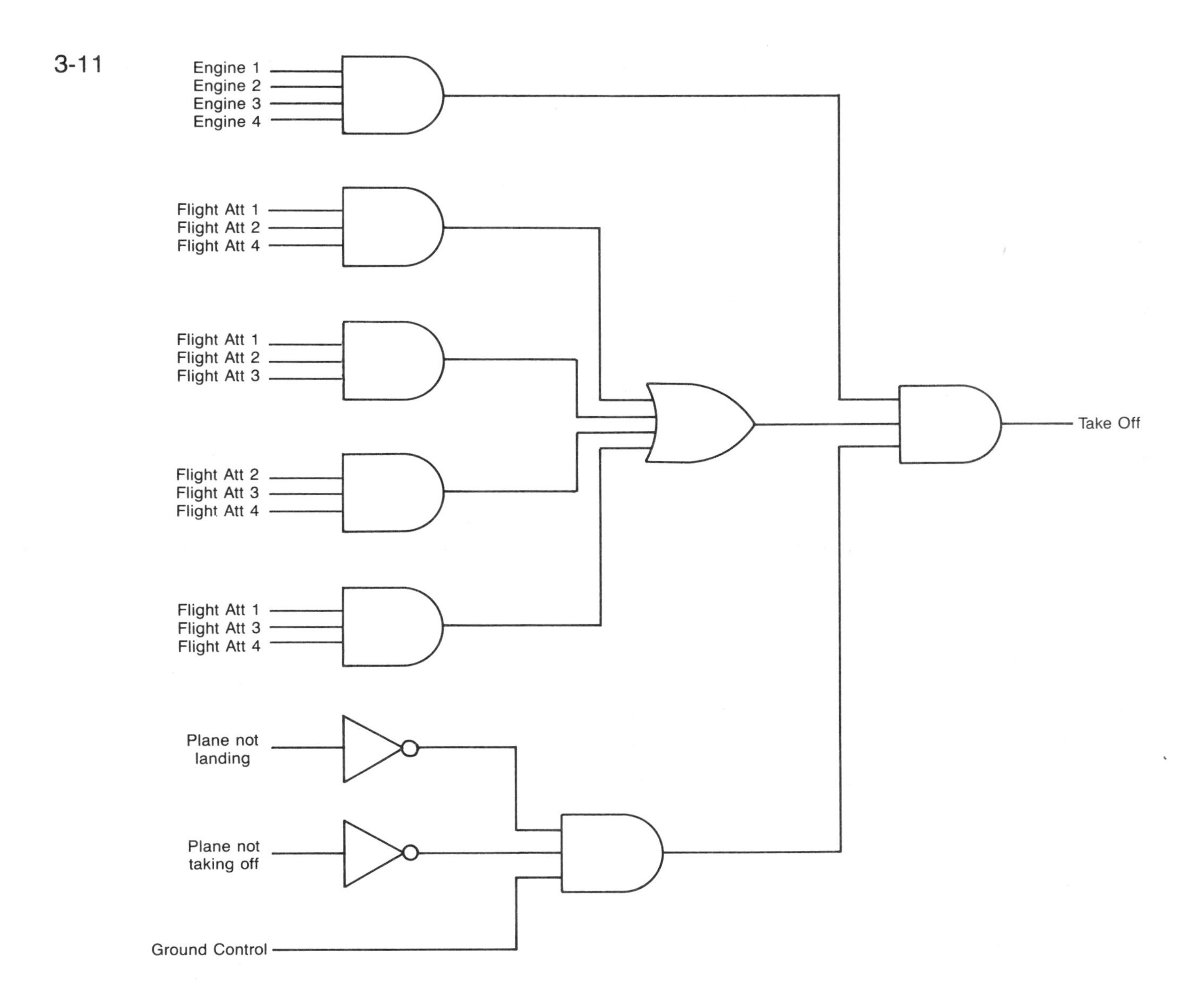

3-12 False. The binary concept has been applied since the invention of Boolean algebra, which has to do with an outcome result based on the true/false (binary) state of the variables.

3-13 b- In series

3-14 A rung is a ladder program term that refers to the programmable contact symbology instructions that drive or control one output.

3-15

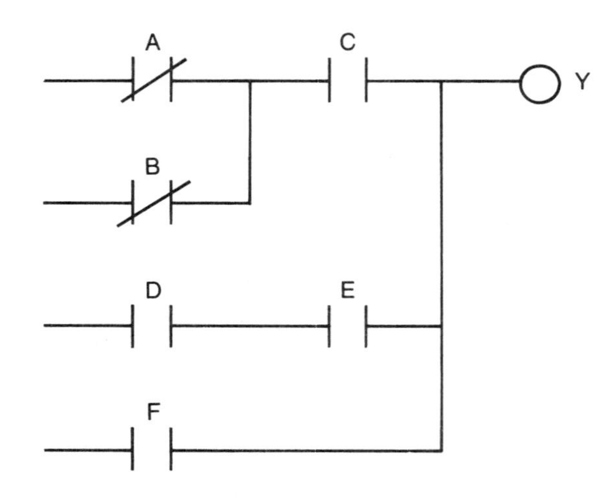

3-16 a and b. Output coil instructions can be used to drive or control real field devices or internal coils. The contacts from the output coils can also be used internally as normally closed or normally open input conditions in a rung.

3-17

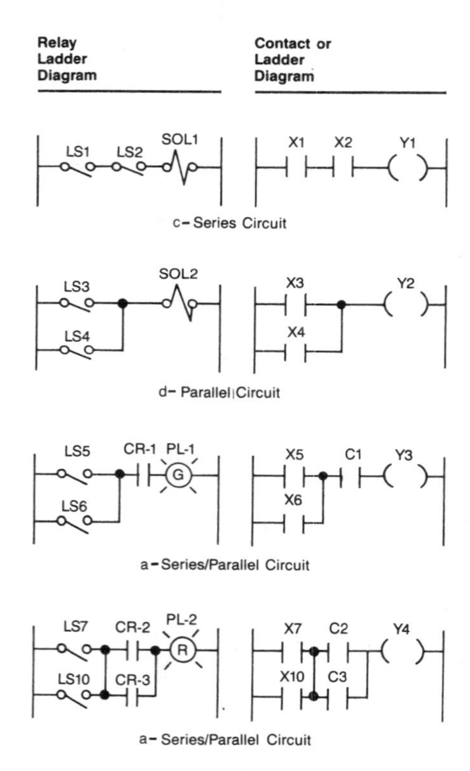

c– Series Circuit

d– Parallel Circuit

a–Series/Parallel Circuit

a– Series/Parallel Circuit

3-17 Continued

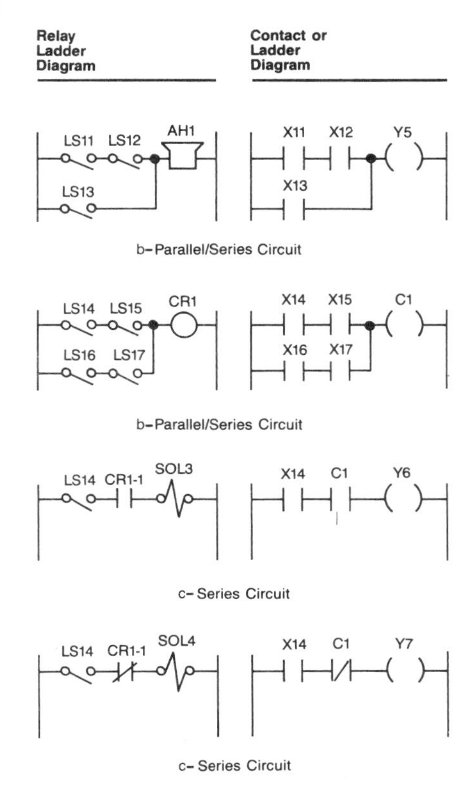

3-18 False. The Boolean algebra was developed by George Boole in the mid-1800s.

3-19 b- At the far right

3-20 False

3-21 A, B → Y

Truth Table

A	B	Y
0	0	1
0	1	0
1	0	0
1	1	0

3-22 NOT output or coil is used to turn the output OFF if all conditions in the rung are true.

3-23 False. In a PC, the only limit to how many contacts can be repeated is a function of the total memory. It is not dependent on a restriction of how many can be used.

3-24 a- Normally closed contacts. A NOT normally open contact will be a normally closed contact.

3-25 False. The maximum number of contact elements in a ladder diagram rung varies according to the programmable controller.

3-26 Internal outputs are used to provide reference for contacts used in other rungs. These internals are generally used for interlocking purposes.

3-27 False. An address can be used as many times as required by the control logic. There is no additional wiring to be done.

3-28

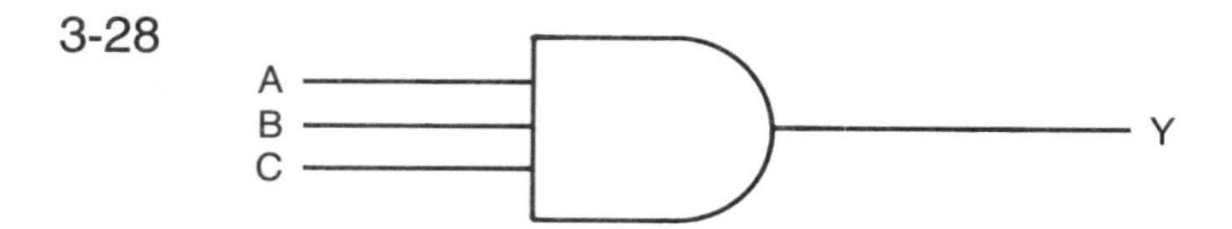

Truth Table

A	B	Y	Y
0	0	0	0
0	0	1	0
0	1	0	0
0	1	1	0
1	0	0	0
1	0	1	0
1	1	0	0
1	1	1	1

3-29 The implementation of the Boolean equation is:

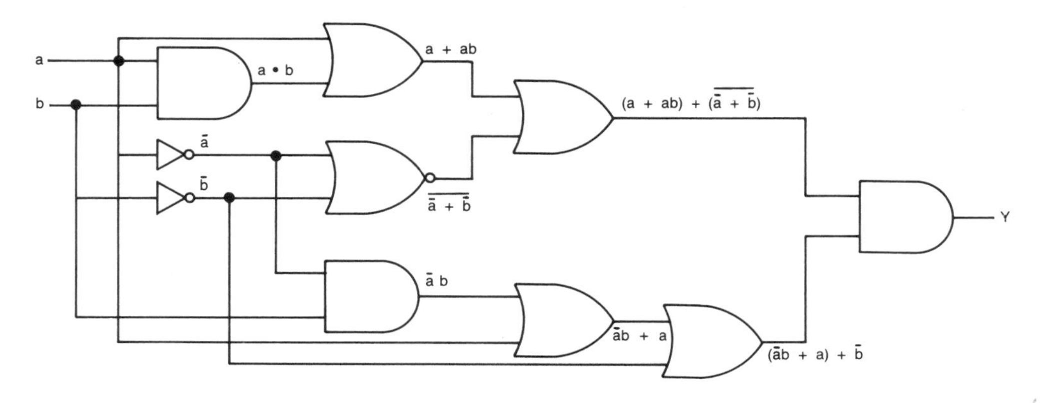

Using Boolean algebra we can minimize the equation as follows:

$[(a + ab) + (\bar{a} + \bar{b})]\ [\bar{b} + (a + \bar{a}b)] = Y$
$[a + ab]\ [\bar{b} + a + b)] = Y$
$[a]\ [\bar{b} + a + b] = Y$
$[a\bar{b} + a + ab] = Y$

The implementation of the minimized equation is:

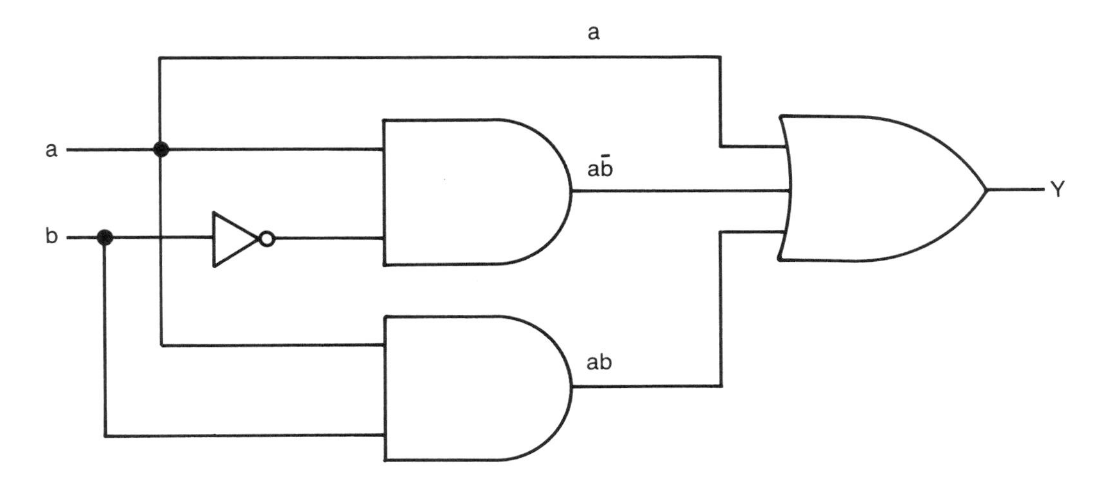

UNIT

--]4[--

ANSWERS

4-1 Processor

4-2 True

4-3 Microprocessor

4-4 Executive

4-5 False. The processor controls not only communication, but also the processing and control of instructions as well as other housekeeping functions.

4-6 Multi-processing

4-7 Independent

4-8 True. The greater the word length capability of the microprocessor, the more bytes of information it can load and manipulate at one time.

4-9 False. The scan time is based on the application memory used.

4-10 Increases

4-11 Program scan time, I/O update time

4-12 Immediate instructions execute an I/O update when evaluated, thus reading an input or writing to an output without finishing the program scan.

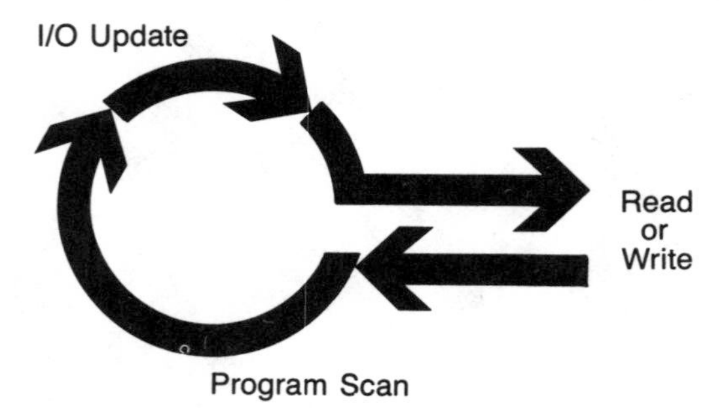

4-13 c- Change state twice in 5 msec

4-14 b- Serial binary format

4-15 Error-checking

4-16 Vertical Redundancy Check (VRC)

4-17 Even or odd

4-18 Parity bit

4-19 For odd parity, the parity bit must be 0; for even parity, the parity bit must be 1.

4-20 c- Single errors

4-21 False. An error would not be detected since the number of ones is still even. Parity can only detect a change in one bit (single error), not in two bits as in this case.

4-22 True

4-23 c- Block check character. The BCC is appended at the end of a block transmission and reflects a characteristic of the block transmitted or checked with the checksum method.

4-24 b- Horizontal redundancy check

4-25 True

4-26 Hamming Code

4-27 Error correcting codes can detect two or more bit errors, but only correct one bit.

4-28 Typical CPU diagnostics include memory, processor, battery, power supply, and communications checks, including I/O subsystem communication diagnostics.

4-29 Watchdog

4-30 c- Executive
d- Scratch pad
a- Application memory
b- Data table

4-31 False. The executive software program is stored in a non-volatile type memory; therefore, no battery back-up is needed.

4-32 Memories can be categorized as volatile and non-volatile.

4-33 Volatile

4-34 Non-volatile

4-35 Read only memory or ROM

4-36 Random access memory or RAM

4-37 Random access memory (RAM) or read/write

4-38 Erasable programmable read only memory or EPROM

4-39 Electrically erasable programmable read only memory or EEPROM

4-40 ROM

4-41 Core memory

4-42 Non-volatile random access memory or NOVRAM

4-43 Core memory is characterized by slow speed, high cost, and large physical space requirements when compared to the integrated circuits provided in RAMs.

4-44 1 or 0

4-45 Bit

4-46 Bit status

4-47 False. Most small PCs have a fixed amount of memory that cannot be expanded.

4-48 1024 or 2^{10}

4-49 c- 65,536 bits capacity. The memory will be 4x1024x16 bits.

4-50 The maximum address will be in decimal 2047 or 2^{11} -1, and in octal the addresses will go from 0 to 3777.

4-51 Memory utilization

4-52 10 rungs x 8 contacts = 80 contacts.
20 rungs x 6 contacts = 120 contacts.
30 rungs x 1 coil = 30 coils.
200 contact total x 3 bytes = 600 bytes.
30 coil outputs x 2 bytes = 60 bytes.
Total Memory 660 bytes.

The total memory required in multiples of 1/8K will be 660 ÷ 128 = 5 20/128; therefore, the requirement will be 6/8K or 3/4K.

With the additional 30% for future expansion, the total memory requirement will be 3/4K x 1.30, or 3.9/4 x K = 7.8/8 K. The total memory requirement will be in multiples of 1/8 K a 8/8 K or simply 1K.

4-53 d- Varies with the controller

4-54 True

4-55 Memory map

4-56 c- The executive and scratch pad are transparent to the user since there is no access to this memory when programming.

4-57 Application

4-58 Data table, user or application program

4-59 True

4-60 True. The input table is an image of the status of field devices that are connected to the PC.

4-61 False. The input table is constantly changing to provide the status of each bit. This change takes place during the I/O update when reading the inputs.

4-62 a- 128 bits. One bit for each output.

4-63 c- 16 bytes, since this is equal to 128 bits, which is the maximum I/O capacity.

4-64 Processor

4-65 The output table is updated during the program scan, while the outputs connected to the field devices are updated during the I/O update (or scan).

4-66 Internal outputs, internal coils, internal control relays, or simply internals.

4-67 Internal outputs are used for interlocking purposes in the control program, just as control relays are used in hardwired logic.

4-68

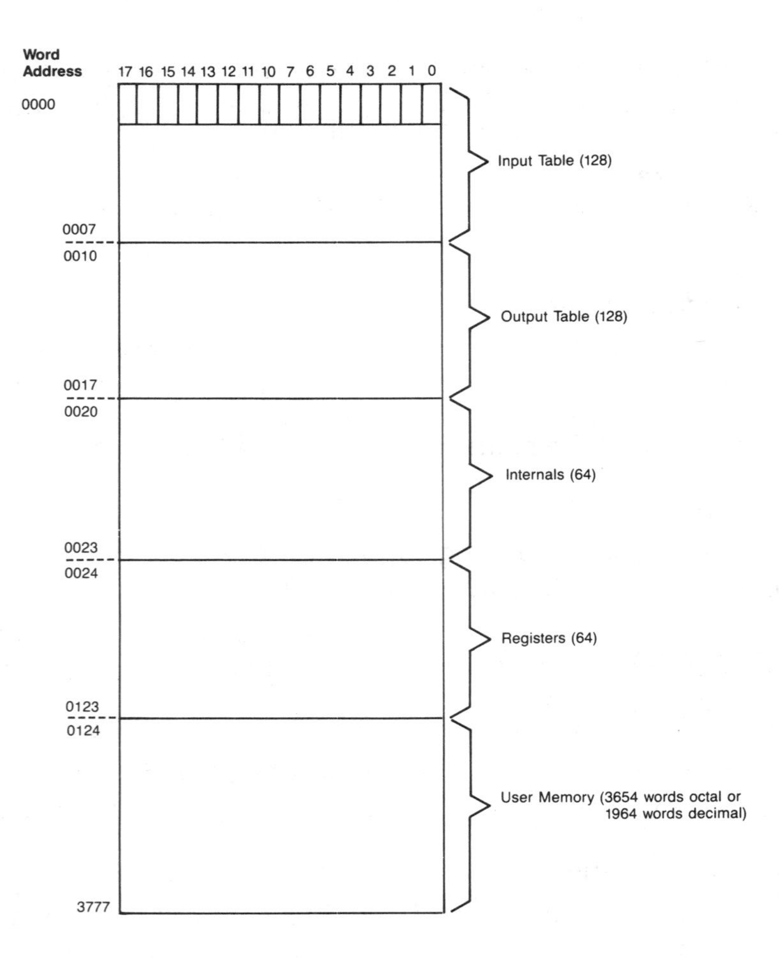

4-69 There are 1964 word locations left, which is 1 235/256 K or approximately 1.918K.

4-70 The data stored in memory could be in BCD format or binary format (including ASCII codes).

4-71 True. The value stored in a register from an analog input will be proportional in value to the range of the analog value.

4-72 C- Timer preset
V- Analog input
V- BCD output
C- Set point
V- Counter accumulated value
C- ASCII message

4-73 True. The addresses of the I/O connected to the field devices are part of the instructions. These addresses are part of the space taken in the memory utilization of the system.

4-74 Executive

4-75 False. The power supply not only provides the necessary voltage levels, but also detects proper rating levels as well as provides protection against possible high-voltage fluctuations.

4-76 False. There are some PCs that can operate with 24 VDC from batteries.

4-77 d- All of the above

4-78 The power supply issues a shutdown command or signal to the microprocessor when the line voltage exceeds the upper or lower limits of a specified voltage level for a specified duration.This duration is generally one to three AC cycles.

4-79 b- Worst-case power loading

4-80 False

4-81 Isolation transformers are generally used when a PC is installed in an environment in which surrounding equipment generates considerable amounts of electro-magnetic interference (EMI) or noise into the power lines.

4-82 False. This problem is very difficult to detect since the overload conditions are a function of the combinations of outputs on at one time, which means that the overload conditions can appear intermittently.

4-83 If the summation of the current requirements for an I/O configuration exceeds the total current supplied by the power supply, a second power supply or a power supply with greater current supply capability may be needed.

4-84 True

UNIT

--]5[--

ANSWERS

5-1 b- Field equipment and the CPU

5-2 d- Discrete inputs/outputs (I/O)

5-3 d- All of the above

5-4 Discrete Input Devices:

Selector switches
Pushbuttons
Proximity switches
Photoelectric sensors
Limit switches
Level switches

Discrete Output Devices:

Motor starters
Solenoids
Valves
Pilot lights
Control relays
Alarms

5-5 Discrete Input Ratings:

24 Volts AC/DC
48 Volts AC/DC
120 Volts AC/DC
230 Volts AC/DC
TTL level (5 Volts DC)

Discrete Output Ratings:

12-48 Volts AC
120 Volts AC
120 Volts DC
230 Volts AC
TTL level (5 Volts DC)

5-6 Logic 1 indicates ON or CLOSED; logic 0 indicates OFF or OPEN.

5-7 Power section and logic section

5-8 c- Isolation circuit. This isolation is between the power and logic sections and usually is included, but not always.

5-9 a- A DC level. The AC/DC input module converts the input signal to a DC level of approximately the same voltage level.

5-10 Typical delays range from 9 to 25 msec due to the filter circuit in the input module.

5-11 c- Exceeds the threshold voltage and remains there for at least the filter delay.

5-12 Isolation is generally provided by an optical-coupler or pulse transformer.

5-13 OFF

5-14 ON

5-15 Solid-state controls with TTL levels or sensing instruments with TTL voltages.

5-16 The input delay caused by filtering is much shorter for TTL inputs than AC/DC inputs.

5-17 c- Closures of dry contact inputs

5-18

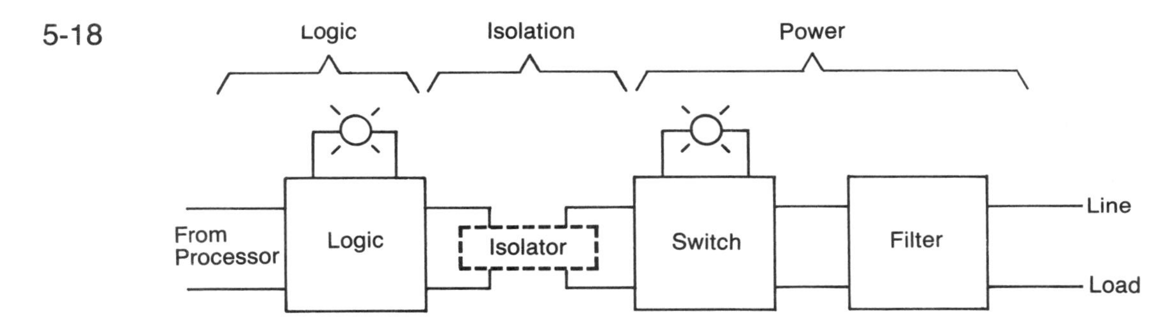

5-19 Triac or SCR (Silicon Controlled Rectifier)

5-20 MOV (Metal oxide varistor) and RC snubber

5-21 Power transistors protected by a free-wheeling diode

5-22 Contact output modules are generally used for switching or multiplexing reference analog signals to drives, switching small currents at low voltages, switching independent source or reference AC voltages with different or internal commons, and contact outputs (with higher ratings) for switching high currents.

5-23 Seven-segment LEDs

5-24 Isolated inputs and outputs may be needed if it is required to connect an input or output device of different ground levels to the controller. Sometimes they are also used when isolation of grounds is required.

5-25

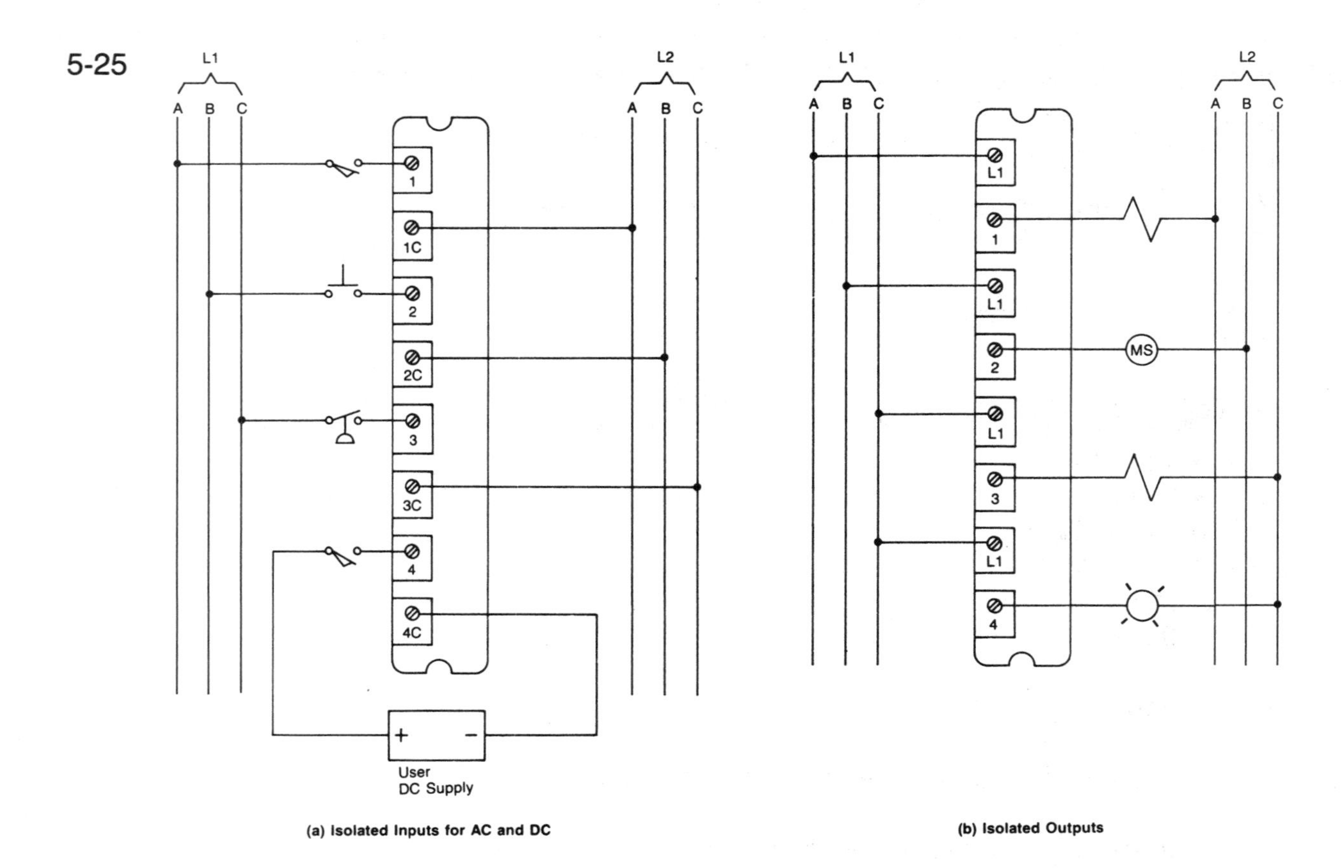

(a) Isolated Inputs for AC and DC

(b) Isolated Outputs

5-26 Data manipulation and arithmetic operation capabilities led to the need for numerical data I/O.

5-27 Multibit discrete I/O and analog interfaces

5-28 AI Pressure transducer
AO Chart recorders
MI Encoders
AI Potentiometers
MI Thumbwheel switches
MO Seven-segment displays

5-29 Unipolar or bipolar type ratings

5-30 Analog input ratings:

0 to +1 Volts DC
0 to +5 Volts DC
0 to +10 Volts DC
—5 to +5 Volts DC
1 to +5 Volts DC
—10 to +10 Volts DC
4 to 20 mA

Analog output ratings

0 to +5 Volts DC
0 to +10 Volts DC
—5 to +5 Volts DC
—10 to +10 Volts DC
4 to 20 mA
10 to 50 mA

5-31 Proportional

5-32 High impedance, high source-resistance outputs

5-33 Shielded cables, located at the PC chassis

5-34 c- 0000 to 9999 BCD or 0 to 32767 decimal. Some controllers use BCD while others may use decimal representations. However, the PCs that use decimal may go up to 2047 or 4095 counts decimal as opposed to 32767.

5-35 Parallel communication between input devices and the processor.

5-36 The most commonly used are thumbwheel switches (TWS), which are used to input parameters such as counter and timer preset values, target set points, and a variety of operator or product codes.

5-37 Range of 5 to 24 volts DC; modules containing 16 or 32 inputs.

5-38 Some output devices include seven-segment displays, BCD alphanumeric displays, and small DC loads. Applications in which they are used include the display or indication of pressure, time, temperature, batch count, and others.

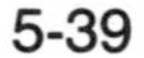

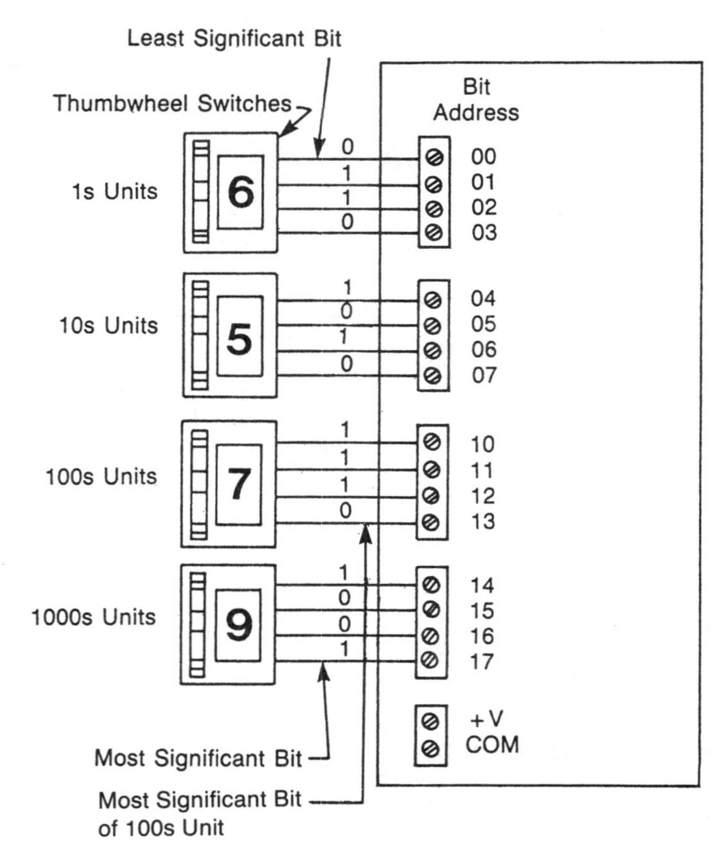

5-40 c- High-speed counters external to the processor

5-41 Encoder input modules are used in applications of closed-loop positioning of machine tool axes, hoists, conveyors, and cycle monitoring of high-speed machines such as can-making equipment.

5-42 Incremental encoders provide pulses that signify position as the encoder rotates; these pulses are received and counted by the encoder module, which in turn sends the information to the processor. Absolute encoders provide BCD or Gray code data, which represent the angular position of the shaft.

5-43 e- Reverse count

5-44 Preprocessing modules

5-45 Microprocessor

5-46 Medium to very large size PCs

5-47 The thermocouple type J input module. These interfaces provide cold junction compensation to correct for changes in cold junction temperature. The operation is similar to an analog input, with the exception that the signals from the thermocouple are connected directly to the module. The signals are very low-level voltage.

5-48

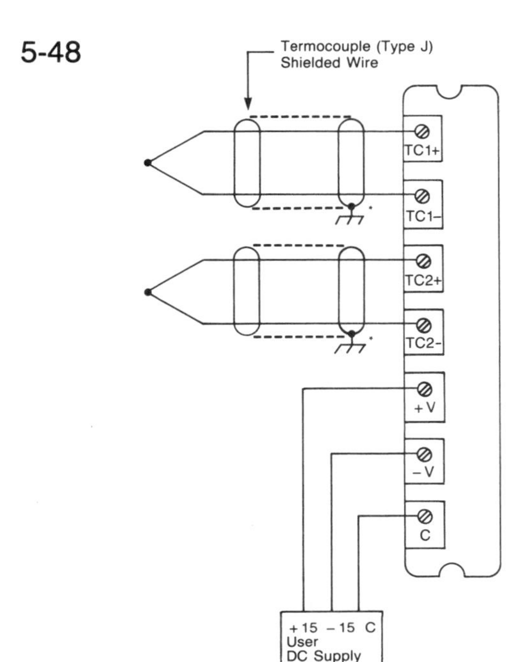

5-49 Pulse stretcher. The fast input module receives very fast input pulses that the module latches so that it can be read in the next scan.

5-50 b- 110 volts AC. The pulse can only be DC voltage.

5-51 Proximity switches, photoelectric cells, or instrumentation equipment which provides pulses in the range of 50 to 100 microseconds.

5-52 The ASCII interface is an input/output module that is used for sending and receiving alphanumeric data between peripheral equipment and a programmable controller. Typical devices that are interfaced with ASCII modules include printers, video terminals, ASCII displays, and computers.

5-53 c- Parallel

5-54 Faster

5-55 c- Every two characters. In the non-smart ASCII module, the characters are sent/received one at a time under interrupt basis. One character is represented in one byte.

5-56 Total scan time = 3.5 msec/K.
Total application program memory used is 4K.
The total scan time will be:

4K x 3.5 msec/K = 14 msec

Based on the communication speed that can be selected, the characters can be sent/ received at the following speeds:

Baud Rate	Character/sec	Speed
300	30	1/30 = 33.3 msec/char.
600	60	1/60 = 16.6 msec/char.
1200	120	1/120 = 8.3 msec/char.
1500	150	1/150 = 6.6 msec/char.
2400	240	1/240 = 4.1 msec/char.

The scan time is 14 msec; therefore, the maximum speed of transmission would be one character every 14 msec or an equivalent baud rate of 714.285. Since this baud is not standard (not available), the maximum speed would be 600 baud.

5-57 If 2 msec more are added to the scan, the total scan would be 16 msec. The maximum speed of transmission would still be 600 baud.

5-58 When setting up ASCII communications, one must consider the matching of baud rates between the sending and receiving equipment, the number of start and stop bits, and the parity (odd, even, or none). Another important detail should be the distance between the ASCII module and the peripheral if RS-232C is used (less than 50 feet).

5-59 c- Have strain gage circuits that provide low-level output signals.

5-60 d- a and b. Strain gage modules are used with pressure transducers and load cells.

5-61 Output module that is used to generate pulse trains compatible with stepper motor translators.

5-62 d- All of the above

5-63 Rate

5-64 d- a and c

5-65 Shorter positioning time, higher accuracy, better reliability, and improved repeatability.

5-66 Servo interface modules combine the feedback pulses from a resolver/encoder to generate an error voltage that is used through a D/A converter to drive a DC motor. The encoder receives the quadrature (cosine wave) phase, which is decoded into forward or reverse count direction and is compared with the drive count (speed).

5-67 Servo position control

5-68 c- Tree-mode closed-loop control

5-69 The basic function of PID is to maintain certain process characteristics at desired set points.

5-70 Process variables may include liquid level, temperature, flow rate, and pressure.

5-71 Kp is the proportional gain.
Ki is the integral gain represented by Kp/Ti where Ti is the reset time.
Kd is the derivative gain represented by KpTd where Td is the rate time.
PV is the process variable.
E is the error, which is equal to PV-SP where SP is the set point.
Vout is the control variable output from the controller.

The proportional gain (Kp) function provides control action that is proportional to the instantaneous error value. The integral gain (Ki) provides additional compensation to the control action, causing a change proportional to the value of the error over a period of time. The derivative gain (Kd) function adds compensation to the control action, causing a change proportional to the rate of change of the error.

5-72 b- Quantity compared to the error signal
c- Desired output
a- Rate or period of update
d- Linearized scaled output
f- Reset action
g- Rate time

5-73

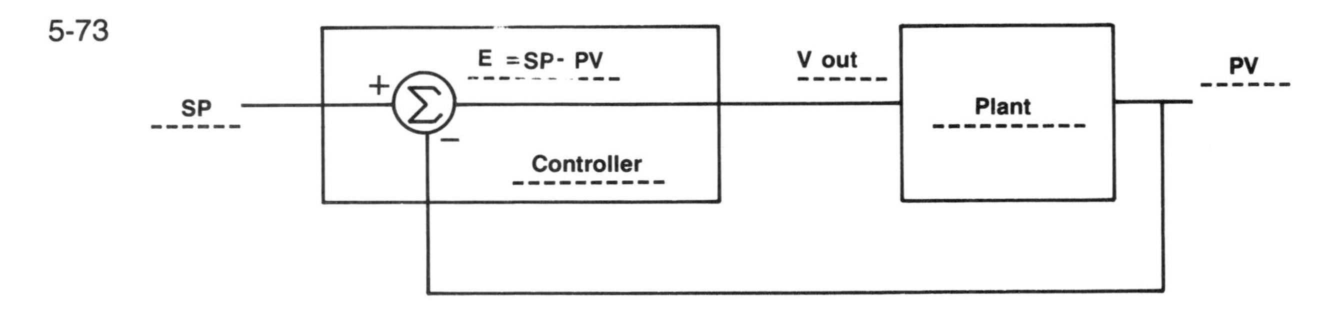

5-74 Data processing modules are used to perform data processing and file handling functions that would typically be performed by the main CPU or dedicated computer. These modules generally are used for the storage and retrieval of recipes, storage and display of messages, report generation, and other processing functions.

5-75 Network interface modules are designed to allow several PCs and other devices to communicate over a high-speed local area communication network. When a message is sent by the processor, the resident network module transmits the message over the network and is received by another network interface in the receiving PC where the information is interpreted and used.

5-76 Typical configurations used in remote subsystems are the daisy chain and star.

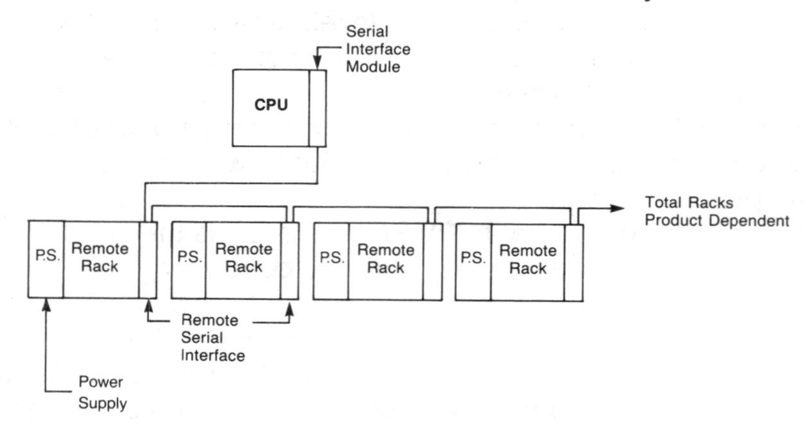

Daisy Chain Configuration

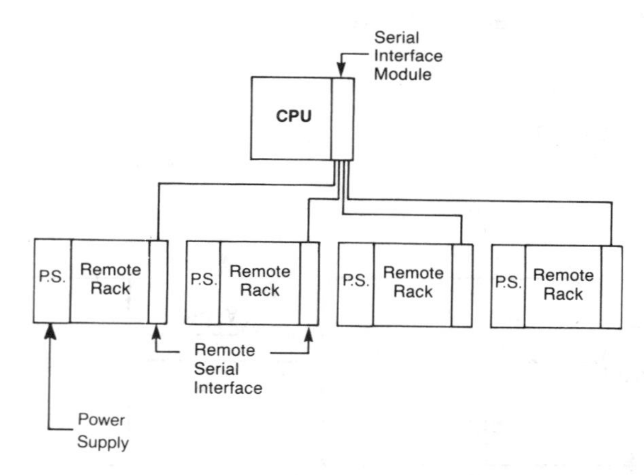

Star Configuration

5-77 Remote subsystems offer large savings on wiring materials and labor costs for large systems in which the field devices are at various spread-out locations. The only cable necessary for communication with the remote units is a pair of twisted cables or coaxial between the subsystem and the main CPU, instead of bringing hundreds of wires from the field devices.

5-78 c- Duration at which input signal must be ON after threshold voltage is passed
f- Maximum leakage of output modules when they are OFF
g- Response time for output to go from ON to OFF
i- Number of I/O circuits in a module
d- Maximum current that an output circuit can carry under load.
a- Nominal AC or DC voltage that specifies magnitude of signal accepted.
k- Voltage isolation between logic and power circuits.
b- Voltage level at which input signal is recognized as being ON.
e- Maximum current duration an output module can withstand.
h- Definition of how close the converted analog signal approximates.
j- Maximum operating temperature.

UNIT

--]6[--

ANSWERS

6-1 Cathode ray tubes (CRTs), miniprogrammers, program loaders, memory burners, and computers.

6-2 True

6-3 CRTs offer the advantage of displaying large amounts of logic on the screen. This logic is shown in the ladder logic format which greatly simplifies the interpretation of the program. CRTs can also be used as display terminals so that information about the process or machine can be available to the operator.

6-4 d- All of the above

6-5 a- On-line, which means that the CRT terminal must be connected to the PC to program the controller.

6-6 d- All of the above. Smart CRTs are intelligent since they have on-board microprocessors with memory and can be used for on-line as well as off-line programming.

6-7 On-line programming implies that the programming terminal must be connected to the PC to create the application program. In this mode, the CPU is in constant communication with the terminal; all the "smarts" associated with the creation of the application program are resident in the PC.

With off-line programming, the terminal does not need to be connected to the PC. The application program can be created with the terminal alone, since all the "smarts" are in the CRT. If a programming device can perform off-line programming, it will also be capable of on-line program creation.

6-8 Miniprogrammers are used for the creation or programming of the control or application software (generally in small PCs). They are also used for monitoring the control logic as well as a debugging tool during start-up.

6-9 False. There are some miniprogrammers that are microprocessor-based. An intelligent mini-programmer (with a micro) is an advantage since the overhead communication time between it and the PC is reduced.

6-10 Program loaders are devices that are used for loading and reloading the control program into the programmable controller memory. Two common types are cassette recorders and electronic memory modules.

6-11 Blast

6-12 Some peripheral device categories used for communication and control with a PC include data entry, documentation, reporting, and displays.

6-13 Thumbwheel switches (TWS) are typically used to enter or modify preset values for timers, counters, shift registers, or other values such as high and low-set point limits.

6-14 b- 12 wires. Each TWS digit requires four wires to transmit the data. Another wire is common to all digits and is used for the voltage source or for the common connection depending on the register input module used.

6-15 BCD format

6-16 f- a and c. A numerical entry panel is used for data entry and as an operator control panel.

6-17 Line printers are used to provide a hard copy printout of the control program. Printers are also used to generate reports or operator messages under program control for data logging purposes.

6-18 Seven-segment displays are interfaced in general with register output modules. Since the format of communication is BCD, it requires four wires per digit to send the information for display.

6-19 Intelligent alphanumeric displays are devices that store canned messages in plain English. The messages are displayed under program control and are used to alert the operator on parameters of the control process.

6-20 b- Used for peripheral documentation

6-21 Manual control stations are devices that provide a manual back-up override of analog or digital-controlled field devices. These control stations are very useful during start-up, override of analog outputs, and back-up of outputs in case of failure.

6-22 P- EIA RS-232C
D- Unibus
P- IEEE 488
D- 20 mA current loop
P- EIA RS-422

6-23 ASCII

6-24 d- All of the above

6-25 Half-duplex

6-26 Full-duplex

6-27 25 possible signal lines

6-28 Some of the electrical characteristics of RS-232C include:

1- The signal voltages at interface point are a minimum of +5V and maximum of +15V for logic 0, and for logic 1, a minimum of -5V and maximum of -15V.

2- The maximum recommended distance is 50 feet or 15 meters; however, longer distances are permissible, provided that the resulting load capacitance measured at the interface point and including the signal terminator does not exceed 2500 picofarads.

3- The drivers used must be able to withstand open or short circuits between pins in the interface.

4- The load impedance at the terminator side must be between 3000 and 7000 ohms, with no more than 2500 picofarads capacitance.

5- Voltages under -3V (logic 1) are called MARK potentials (signal condition); voltages above +3V (logic 0) are called SPACE voltages. The area between -3V and +3V is not defined.

6-29 b- 20 mA current loop. This standard achieves greater distances since the wire resistance has no effect on a constant current loop. In a RS-232C, the voltage drops across the wire, thus limiting a maximum wire length for proper transmission.

6-30 d- There is a checksum bit for error detection. An ASCII character can have an added bit for parity error checking, but not a checksum bit. However, a block of several ASCII characters transmitted can have a check sum character at the end of the block transmission.

6-31 b- Unbalanced link
c- RS-232C
d- Balanced link
a- RS-449
f- Logic 1
g- Logic 0
h- 1 stop bit
e- 2 stop bits

6-32

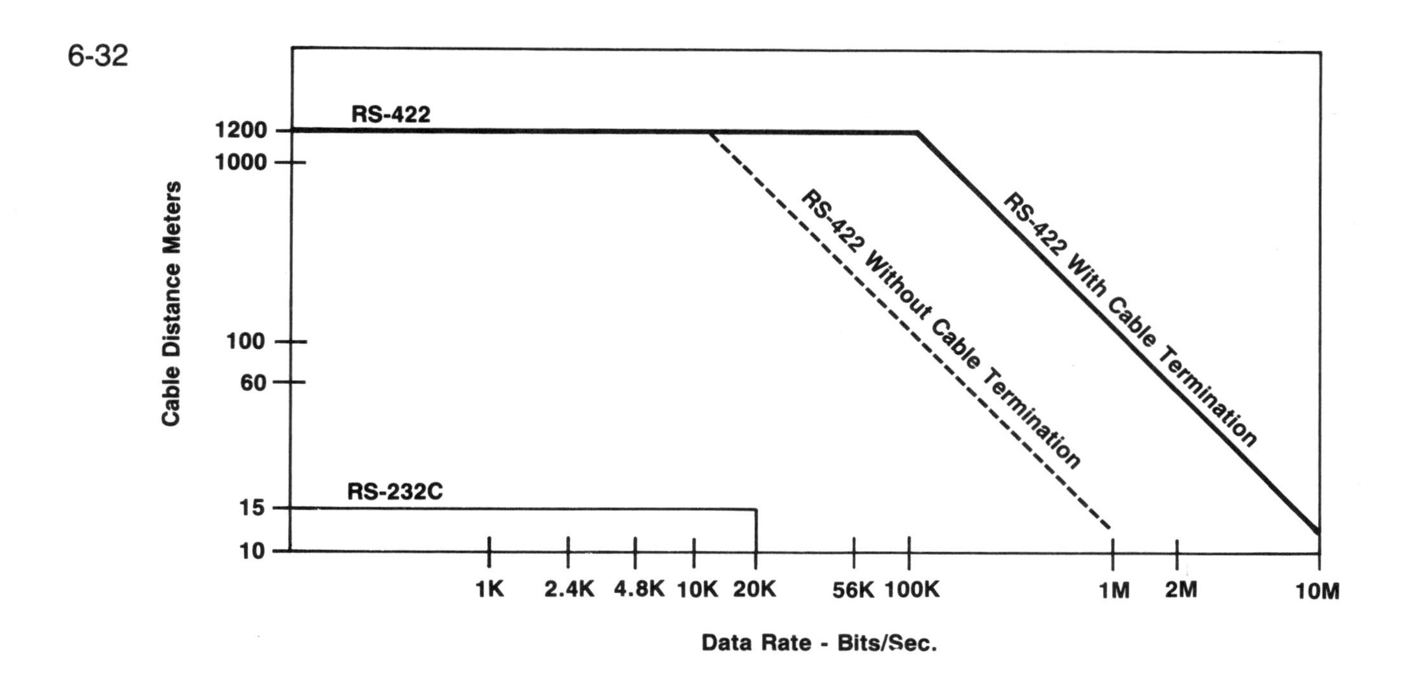

6-33 d- Signal ground

UNIT
--]7[--

ANSWERS

7-1 Ladder language

7-2 True

7-3 It is a set of input conditions represented by contact instructions and an output instruction at the end.

7-4 Contacts and coils

7-5 Relay logic, arithmetic, data transfers, timing and counting, data manipulation, and flow of control

7-6 b- Coil symbols

7-7 Ladder diagrams, Boolean mnemonics, functional blocks, English statements.

7-8 a- Functional blocks
d- English statements

7-9 Basic PC instructions perform relatively easy operations, such as relay replacement, timing, and counting.
High-level instructions perform more powerful operations, such as analog control, data manipulation, and reporting.

7-10 True. However, most PCs will have only one programming language.

7-11 It is a complete continuous path from left to right; moreover, it is having a true condition.

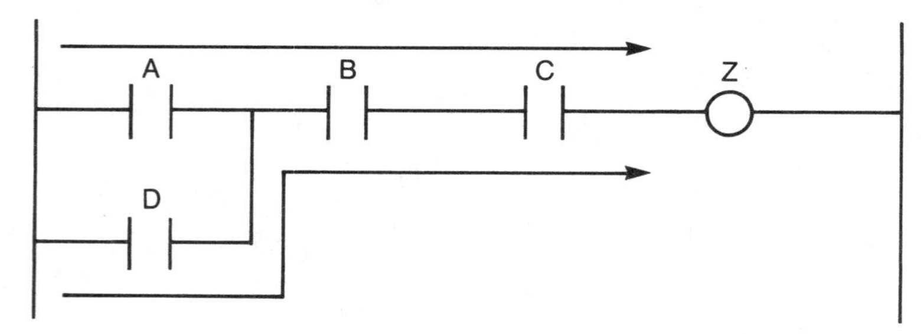

Continuity exists if contacts A, B, and C close, or if contacts D, B, or C close.

7-12 False

7-13 It is used to control either an output connected to the controller or an internal output.

7-14 Using an unlatched coil instruction of the same reference address.

7-15 The contact address must change states at least twice.

7-16 TON: When the accumulated time equals the preset time, the output is energized and the timed-out contact associated with the output is closed. If logic continuity is lost before the timer is timed-out, the accumulated value resets to zero.

TOF: When the accumulated time equals the preset time, the output is de-energized and the timed-out contact associated with the output is opened. If logic continuity is gained before the timer is timedout, the accumulated value resets to zero.

7-17 b- OFF-to-ON transition

7-18 By using a NC contact of the TON coil to drive the desired output coil.

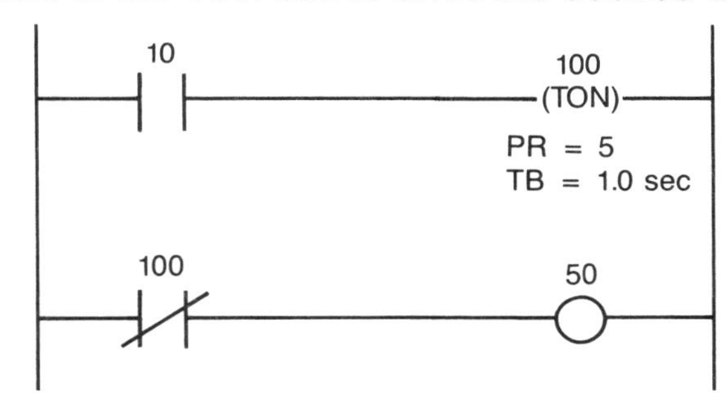

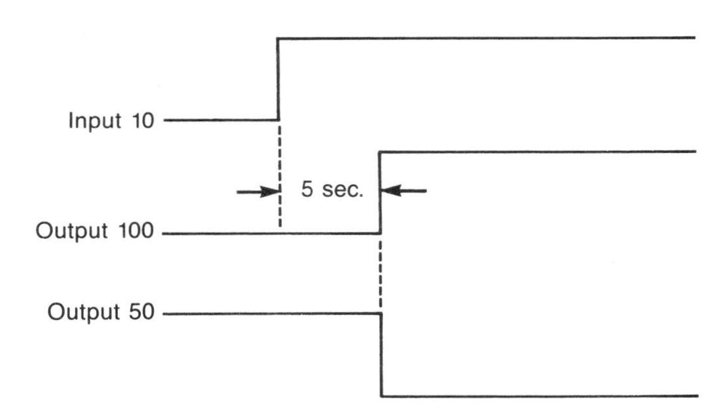

Output 50 will go OFF 5 seconds after A closes.

7-19 It is the only automatic means of resetting the accumulated value of a retentive timer.

7-20 If input 100 closes, the contents of R145 are multiplied with the contents of R130 and stored in R135. Output 50 is ON when the multiplication is enabled.

7-21 The MCR instruction is used to activate or de-activate the execution of a group of ladder rungs. It could be used in set-up conditions as well as in the handling of different parameters.

7-22 When input 10 (SS1) is enabled, the value of the thumb-wheel switches (BCD) is stored in destination register 2000 (BLK IN instruction) and output 30 is energized. The second block instruction reads the analog input signal (in slot 4) and stores the value in register 3000. Output 31 is energized when the block is enabled. The MOV BCD-D (move BCD to Decimal) instruction converts the value stored in register 2000 (BCD) to a decimal value, which is stored in register 2001. Output 32 is energized when the block is enabled, which is always. The compare block (CMP) compares the values of registers 3000 and 2001; if the contents of 3000 are greater or equal to 2001, output 33 turns ON, which in turn energizes output 20 (solenoid 1).

7-23 c- Contacts 10, 11, 8, and 7 CLOSED. For this sequence, contact 8 would be reverse power flow which is not allowed. The following circuit will work for all sequences (a,b,c, and d):

This circuit will work for all the sequences (a, b, c, and d).

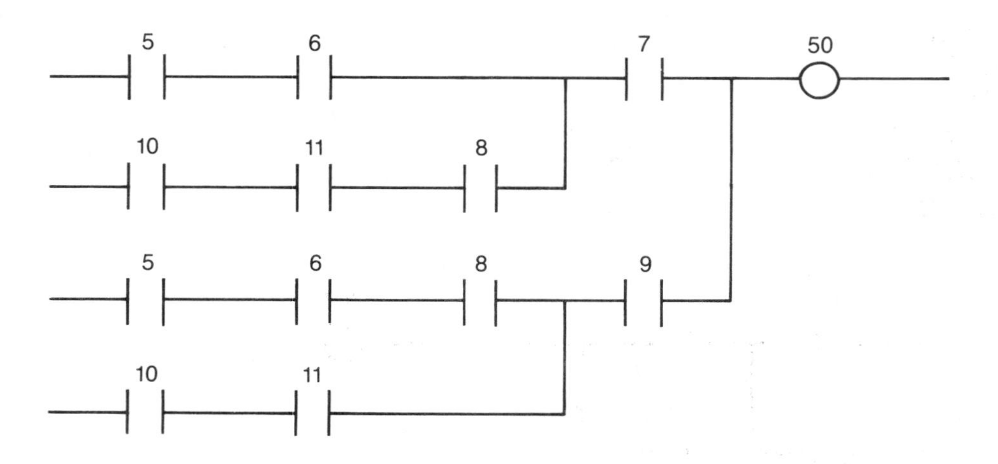

7-24 a- TON delay energize
b- TON delay de-energize

7-25 a- LD 1
AND 2
LD 4
AND 5
OR LD
LD 3
OR NOT 6
AND LD
OUT 100

7-25 Continued

b- LD 1
AND 2
LD
AND 5
OR LD
AND 3
OR NOT 6
OUT 200

c- LD NOT 1
AND 2
LD 5
AND 6
OR LD
AND 3
LD 4
OR NOT 7
AND LD
OUT 250

7-26 a-

b-

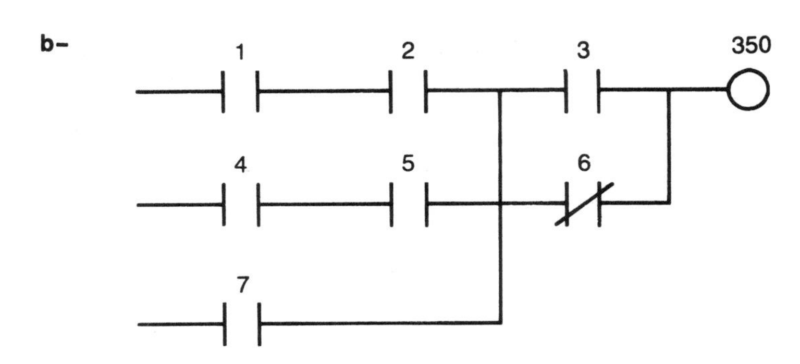

7-26 Continued

c-

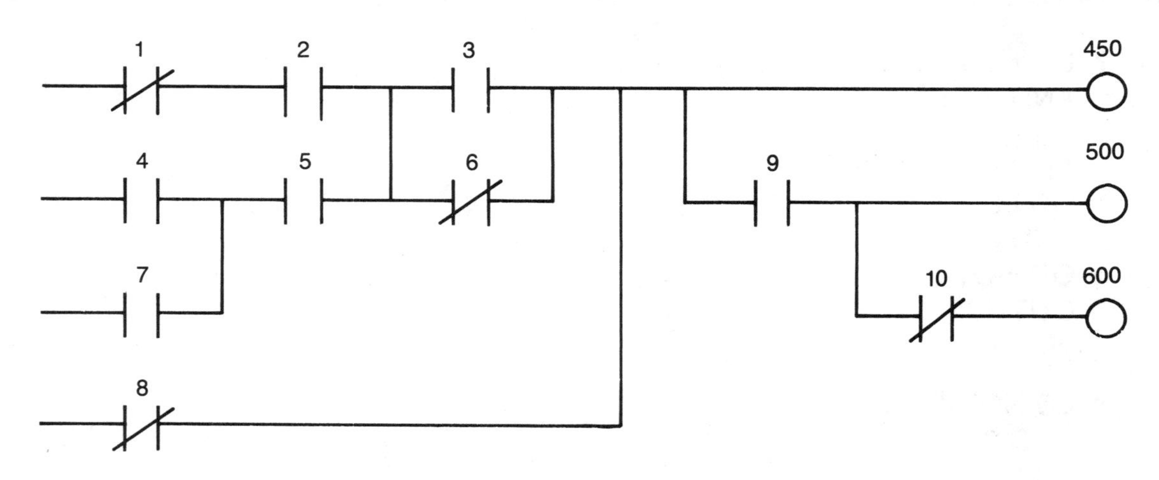

7-27 Interruptable timers, accumulated value

7-28 Edge triggered

7-29 No

7-30 Comparison

7-31 Double precision, two

7-32 False. Only two registers; matrix comparisons could compare two or more registers.

7-33 False. It will be lost in the shift instruction; in the rotate instruction, the least significant bit will be placed in the most significant bit location.

7-34 d- Examine bit

7-35 c- PID

7-36 False. The programming language is part of selecting the controller. The programming language can not be added to an existing PC.

7-37 c- Ladder diagrams

7-38 True. Different types of languages will have different scan times per K of application memory used.

7-39 BCD to binary, binary to BCD, absolute, complement, inversion.

7-40 d- None of the above. Logic matrix instructions are used to perform these operations.

7-41 Move block

7-42 Diagnostic

7-43 True. You could have as many as you want. The number will depend on the type of PC.

7-44 c- ASCII transfer

UNIT
--]8[--

ANSWERS

8-1 c- Define the control task. This is the first step that must be done; without knowing what needs to be controlled nothing can be done.

8-2 Algorithm

8-3 The following guidelines are recommended as an approach for program design in modernization projects:

1- Understand the actual process or machine function.
2- Review machine logic of operation and optimize when possible.
3- Assign real I/O addresses and internal addresses to inputs and outputs.
4- Translate relay ladder diagram to PC coding.

8-4 Relay ladder diagram

8-5 b- Specifications

8-6 Flowcharting

8-7 d- All of the above. When developing a control program, one can use logic gates, relay ladder symbology, or PC contact symbology. The choice depends on what the programmer feels more comfortable with.

8-8

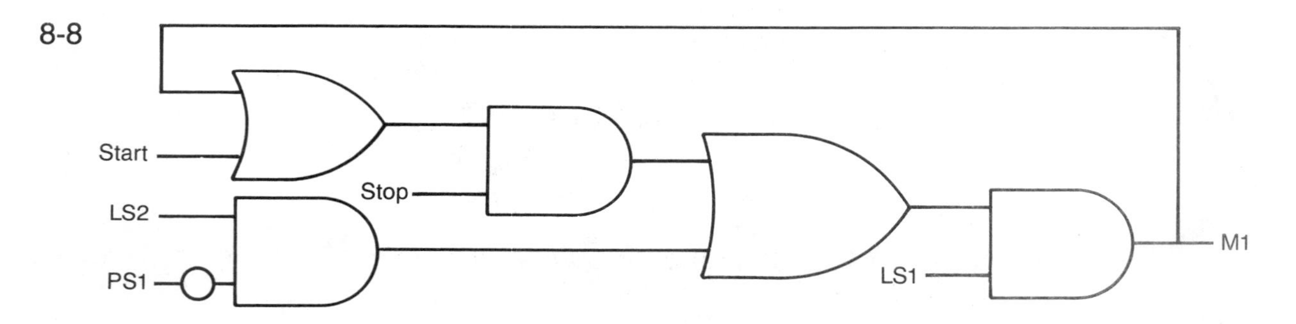

8-9

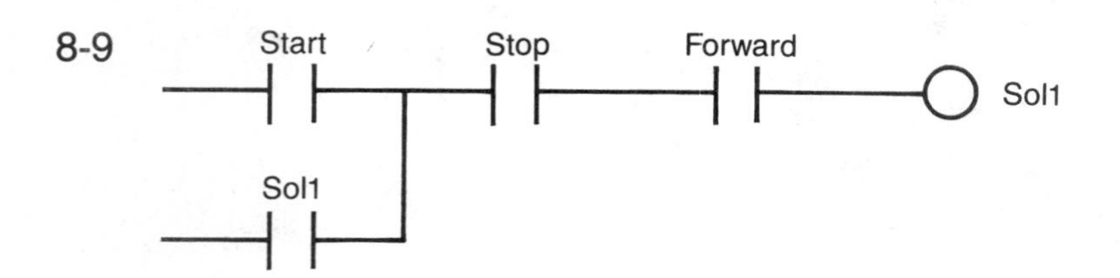

8-10

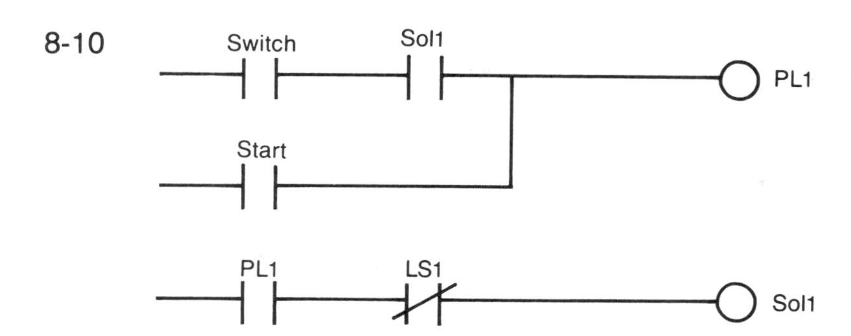

8-11 The address assignment process is one of the most important procedures that takes place during the programming stage. This process indicates what PC inputs are connected to what input field devices, and what outputs will drive what output field devices. The assignment of internals, timers, and counters takes place during this stage. The I/O address will be the reference by which each input and output is referred to in the control program.

8-12 False. Internals, MCRs, timers, and counter addresses are also included in this step.

8-13 The numbers assigned to the I/O addresses depend on the PC model used.

8-14 Octal, decimal, or hexadecimal number systems

8-15

I/O Address			Slot	Module Type	Description
Rack	Group	Terminal			
0	0	0	0	Input	Limit Switch LS10
0	0	1	0	Input	Pressure Switch PS3
0	0	2	0	Input	Motor Contact M11
0	0	3	0	Input	Start PB SPB1
0	0	4	0	Input	Reset PB RPB3
0	0	5	0	Input	— Not Used —
0	0	6	0	↑	
0	0	7	0		
0	1	0	0		
0	1	1	0		
0	1	2	0	Not Used	
0	1	3	0		
0	1	4	0		
0	1	5	0		
0	1	6	0		
0	1	7	0	↓	
0	2	0	1	Output	Solenoid Sol3
0	2	1	1	Output	Motor M11
0	2	2	1	Output	Pilot Light PL4
0	2	3	1	Output	— Not Used —
0	2	4	1	↑	
0	2	5	1		
0	2	6	1	Not Used	
0	2	7	1		
	•				
	•				
	•				
0	3	7	1	↓	

8-16 Only one input needs to be connected to an input module. If this multiple input device is called for in several rungs, it will have the same address, whether normally open or normally closed.

8-17 a-

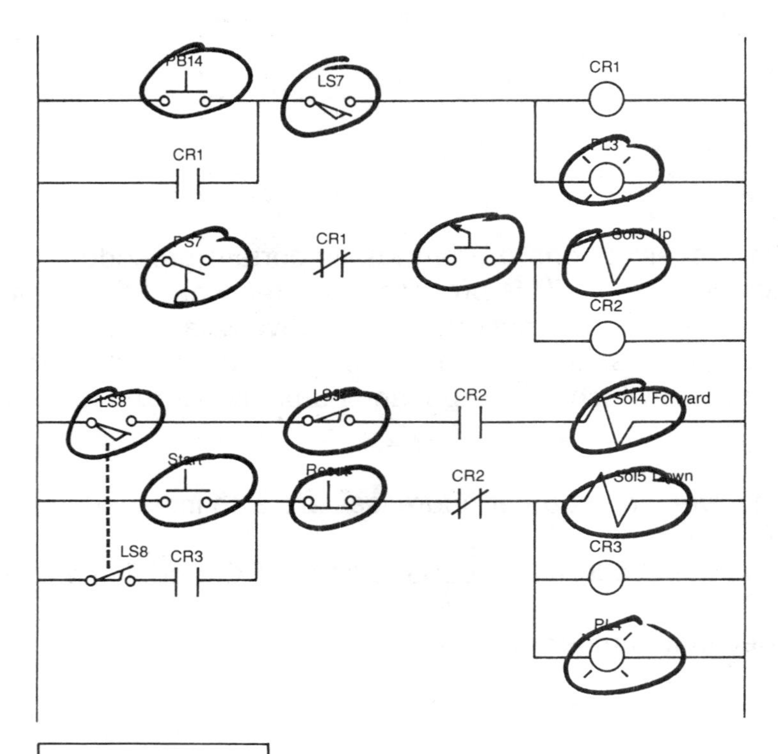

b-

I/O Address				
Rack	Group	Terminal	Module Type	Description
0	0	0	Input	PB14
0	0	1		LS7
0	0	2		PS7
0	0	3		SEL
0	0	4		LS8
0	0	5		LS9
0	0	6		Start PB
0	0	7		Reset PB
0	1	0	Spare	← Not Used →
	• • •			
0	1	7		
0	2	0	Output	PL3
0	2	1		Sol3 Up
0	2	2		Sol4 Forward
0	2	3		Sol5 Down
0	2	4		PL4
0	2	5		— Not Used —
0	2	6		— Not Used —
0	2	7		— Not Used —
0	3	0	Spare	← Not Used →
	• • •			
0	7	7		

8-17 Continued

C-

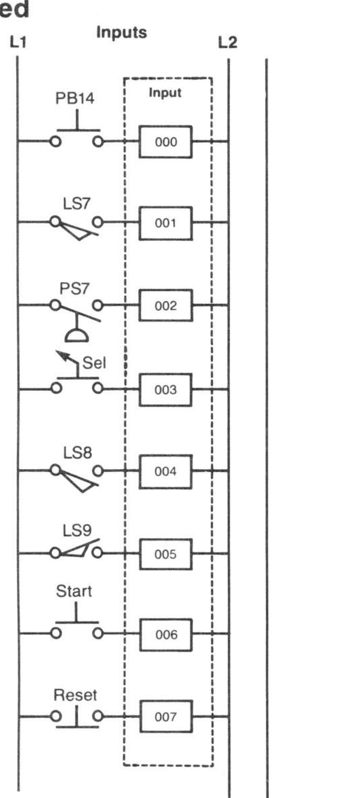

PC Program Coding

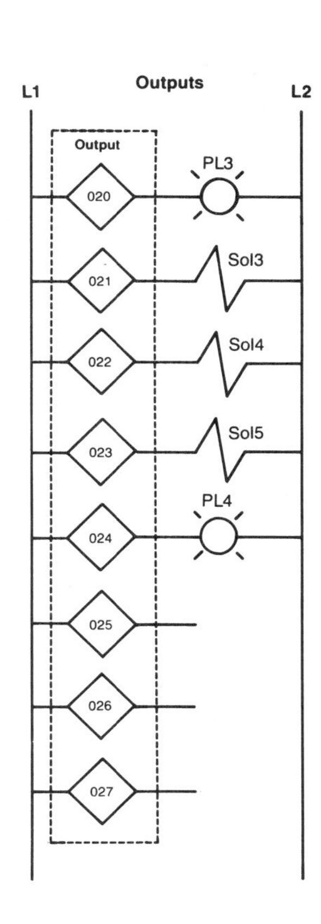

8-18 a-

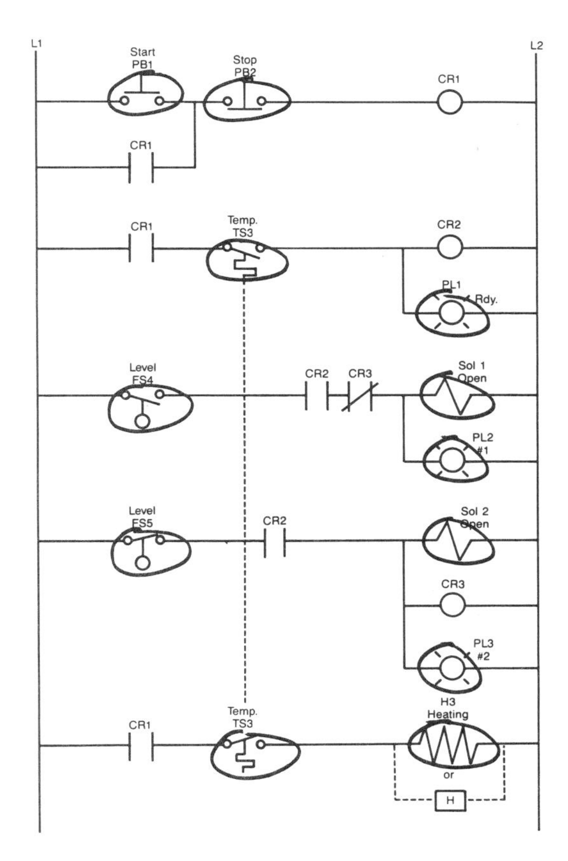

8-18 Continued

b-

I/O Address				
Rack	Group	Terminal	Module Type	Description
0	0	0	Input	Start PB1
0	0	1		Stop PB2
0	0	2		Temp TS3
0	0	3		Level FS4
0	0	4		Level FS5
0	0	5		— Not Used —
0	0	6		— Not Used —
0	0	7		— Not Used —
0	1	0	Spare	← Not Used →
0	1	1		
0	1	7		
0	2	0	Output	PL1 Ready
0	2	1		Sol1 Open
0	2	2		PL2 #1
0	2	3		Sol2 Open
0	2	4		PL3 #2
0	2	5		H3 Heating
0	2	6		
0	2	7		
0	3	0	Spare	← Not Used →
0	3	1		
0	7	7		

c-

Device	Internal	Description
CR1	1000	Control Relay CR1
CR2	—	Same as PL1 Ready
CR3	—	Same as Sol2 Open

8-18 Continued

d-

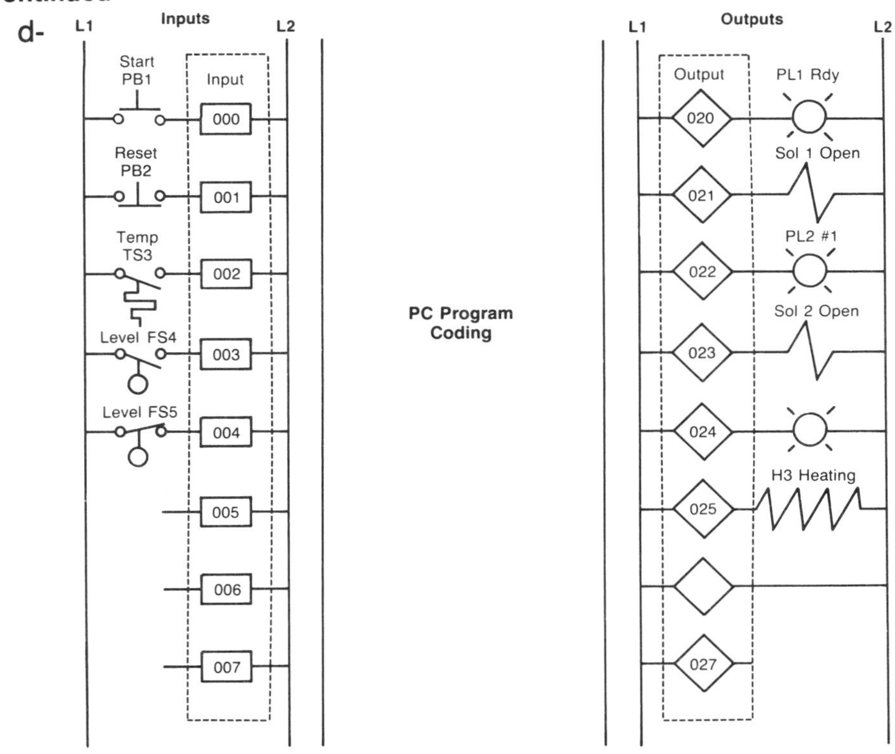

8-19 c- Ensure safety

8-20 b- In series

8-21 d- To shutdown the system if there is a PC failure

8-22 The SCR is improperly connected since the PC will never energize its fault contacts because it cannot have power.

8-23 The NC PC fault contacts are used to energize the PC failure alarm. This alarm will occur if the PC fault coil is not energized, the NO PC fault contacts will not close, and the NC PC fault contacts will remain closed.

8-24 Use a NC contact from a control relay (or from the SCR) as shown below.

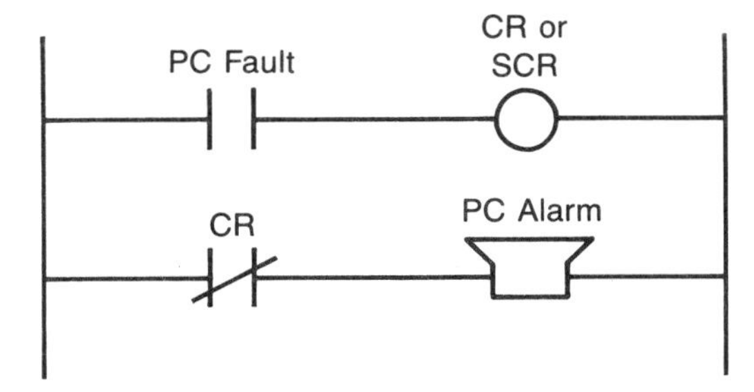

8-25 Coding

8-26 LS1 is wired NO; LS2 is wired NC.

8-27

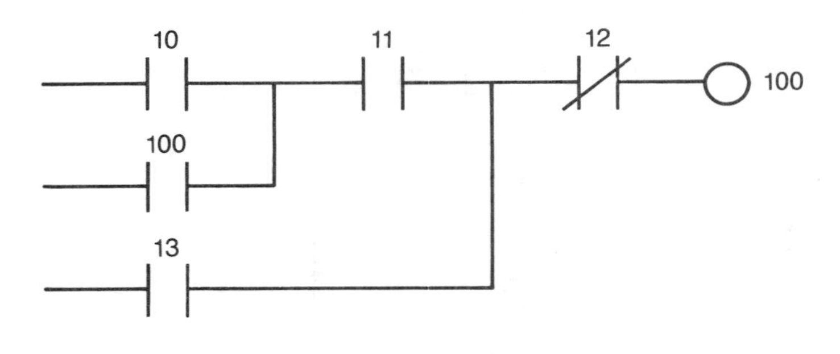

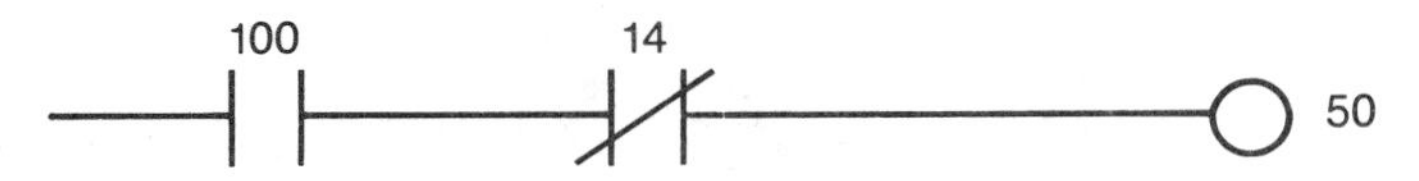

I/O Address	Device	Type
10	Start	Input
11	LS1	Input
12	FS3	Input
13	PS2	Input
50	Sol2	Output
100	I	Internal

8-28

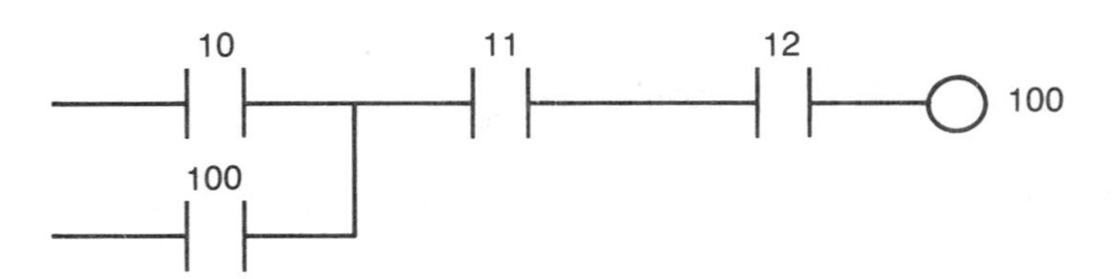

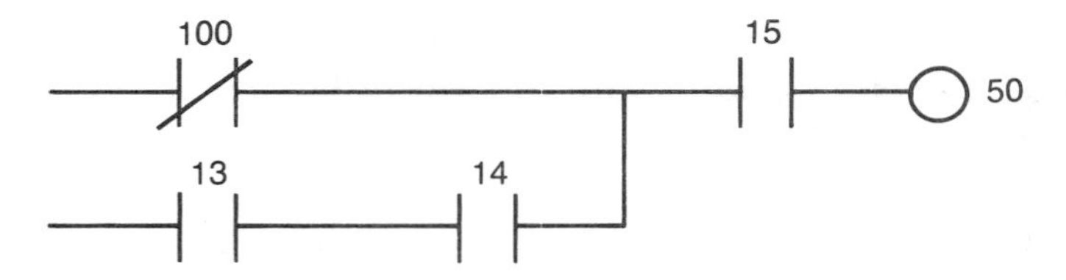

I/O Address	Device	Type
10	LS14	Input
11	SS3	Input
12	PS1	Input
13	S4	Input
14	LS15	Input
15	SS4	Input
50	Sol7	Output
100	CR10	Internal

8-29 False. Although most of the time an NC input device is programmed NO, it depends on the requirement of operation for the input.

8-30 c- Programmed NO
b- NO contact close
a- Programmed NC
d- NO contact open
d- NC contact close
b- NC contact open

8-31 b- The contacts open

8-32 a- Stays in the programmed state which is normally closed.

8-33 True. If the wire connecting the input device to the module is broken, the programmed NO contacts will remain open.

8-34 True. A normally closed device can be programmed normally closed if it would interrupt power flow if the device is not activated (remains closed).

8-35 a-

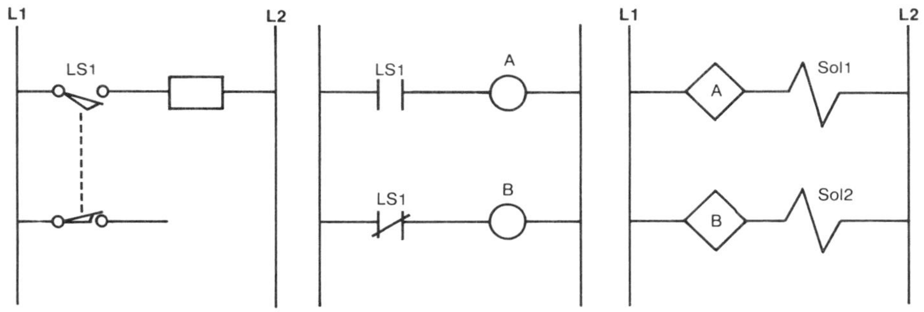

b-

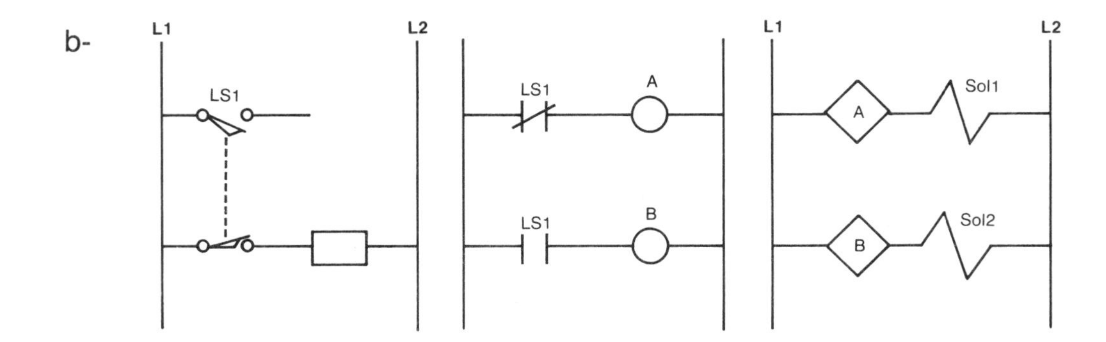

8-36 If a failure occurs and an NC limit switch has to be replaced by an NO limit switch, all the contact addresses in which this limit switch is used must be changed to an NC. When the proper replacement is found, the contact addresses must be changed back to NO.

8-37 Advantages for modernizations include a more reliable system, less energy consumption, less space used for the control enclosure, and a flexible system that can be expanded.

8-38 a- The control relay CR1 does not need an internal since the address is the same as Sol1 or PL1.

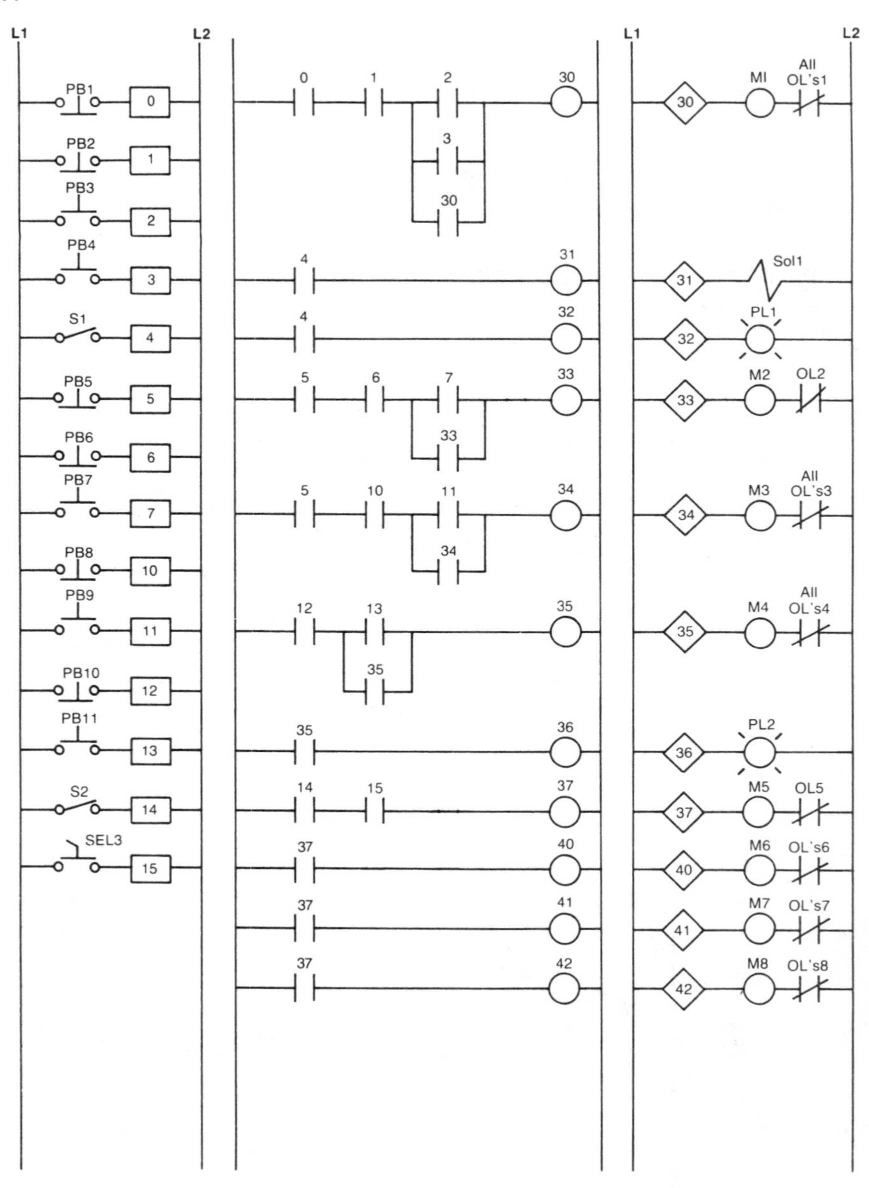

8-38 Continued

b- Same answer as 8-38a except that the contacts with addresses 0, 1, 5, 6, 10 and 12 should be programmed NC.

c- M1 and M2 are assigned addresses 16 and 17 and are used in output rungs 30 and 35. The OL2 and OL5 are assigned inputs 20 and 21 and are used in output rungs 33 and 37.

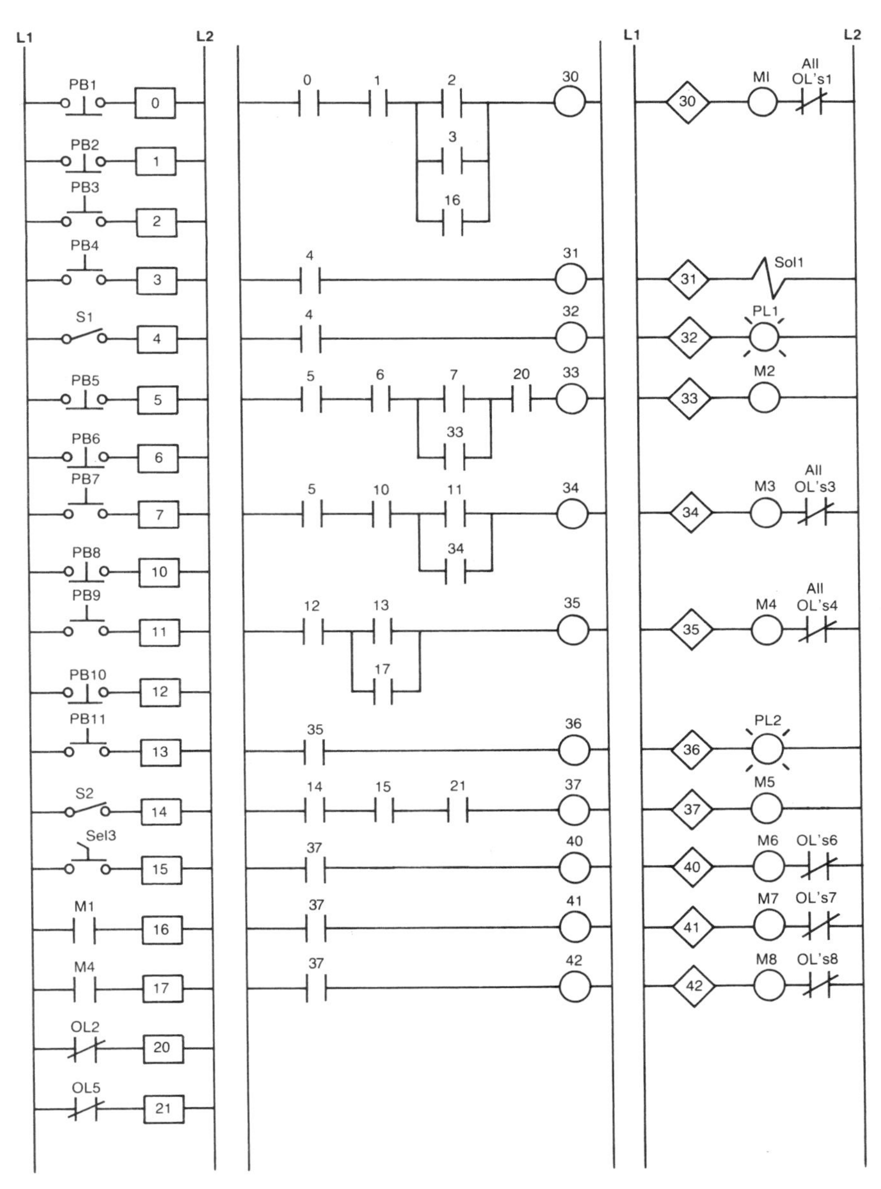

8-39

8-40

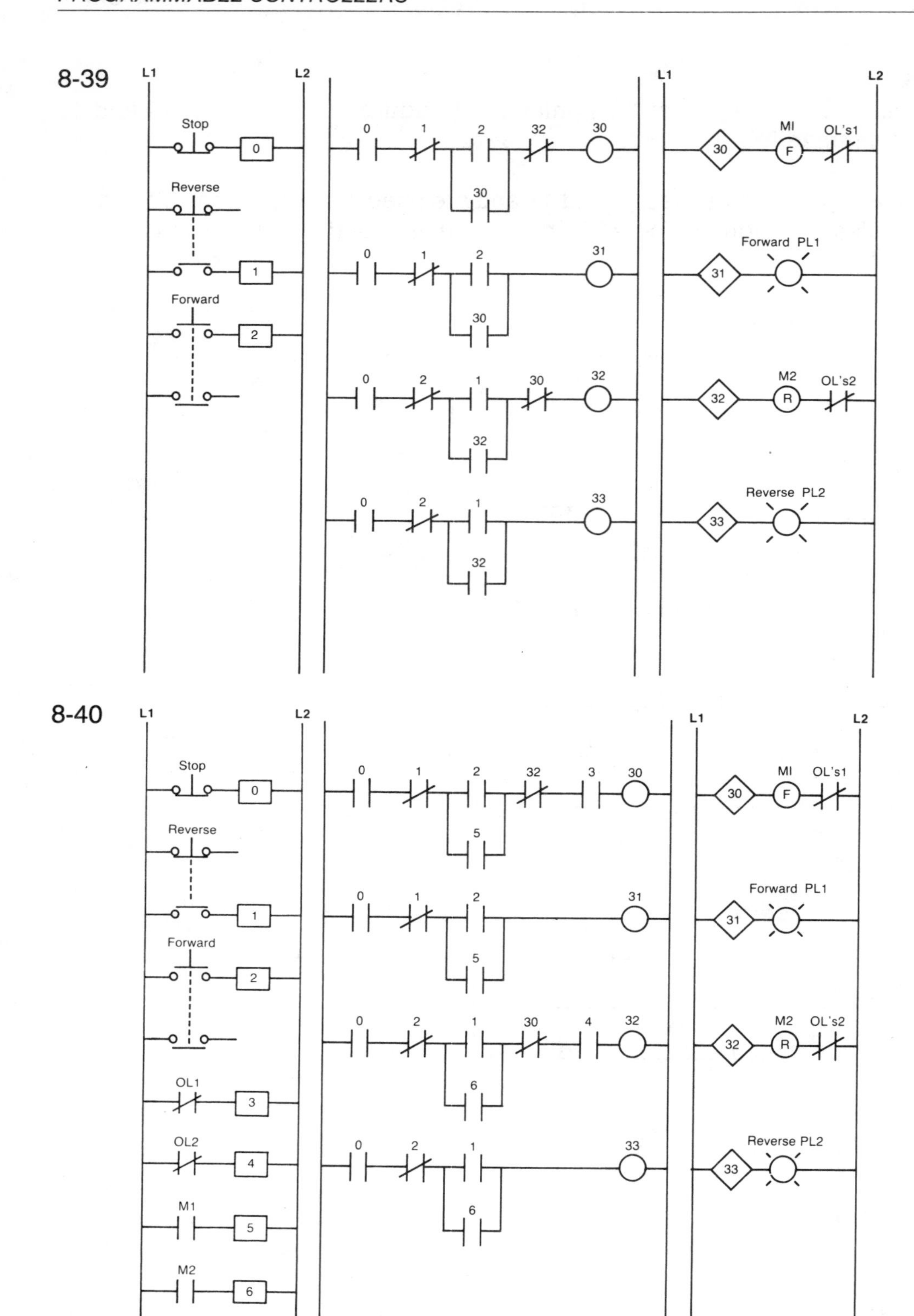

L1
L2
Stop
Reverse
Forward
OL1
OL2
M1
M2
MI
OL's1
Forward PL1
M2
OL's2
Reverse PL2

8-41

Address Assignment:

I/O Address			Type	Description
0	0	0	Input	PB1
0	0	1	Input	PS1
0	0	2	Input	FS1
0	0	3	Input	TS1
0	0	4	Input	LS1
0	0	5	Input	PS2
0	0	6		↑
	•			Not Used
0	2	7		↓
0	3	0	Output	Sol1
0	3	1	Output	Sol2
0	3	2	Output	Sol3
0	3	3		↑
	•			Not Used
0	4	7		↓

Internal Assignment:

Internal Address	Device	Description
100	TMR1	Use to trap TMR1
—	CR1	Same as Sol1
—	CR2	Same as Sol2
101	CR3	Replace CR3
200	TMR1	Timer 1
201	TMR2	Timer 2

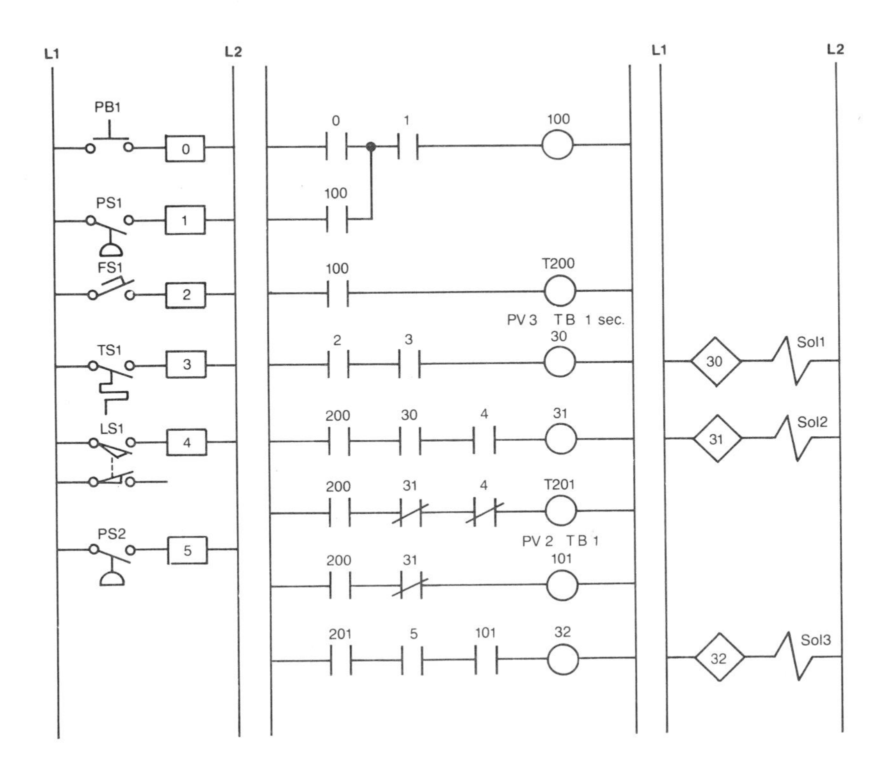

8-42 The analog output will be connected to terminal blocks 2(+) and 3(-) for the speed reference. The forward and reverse signals switch the common through the contact output modules (outputs 21 and 22). The overloads are connected in series to TB1-9 and TB1-8 with the 115 VAC source internal to the drive; output 20 switches the start signal for the drive.

The program coding takes care of the RUN/JOG (input 13) condition; jogging is accomplished by de-energizing input 13 (JOG condition) and pressing the START PB. Outputs 20 and 21 will switch, in a mutually exclusive form, the forward and reverse commands to the drive.

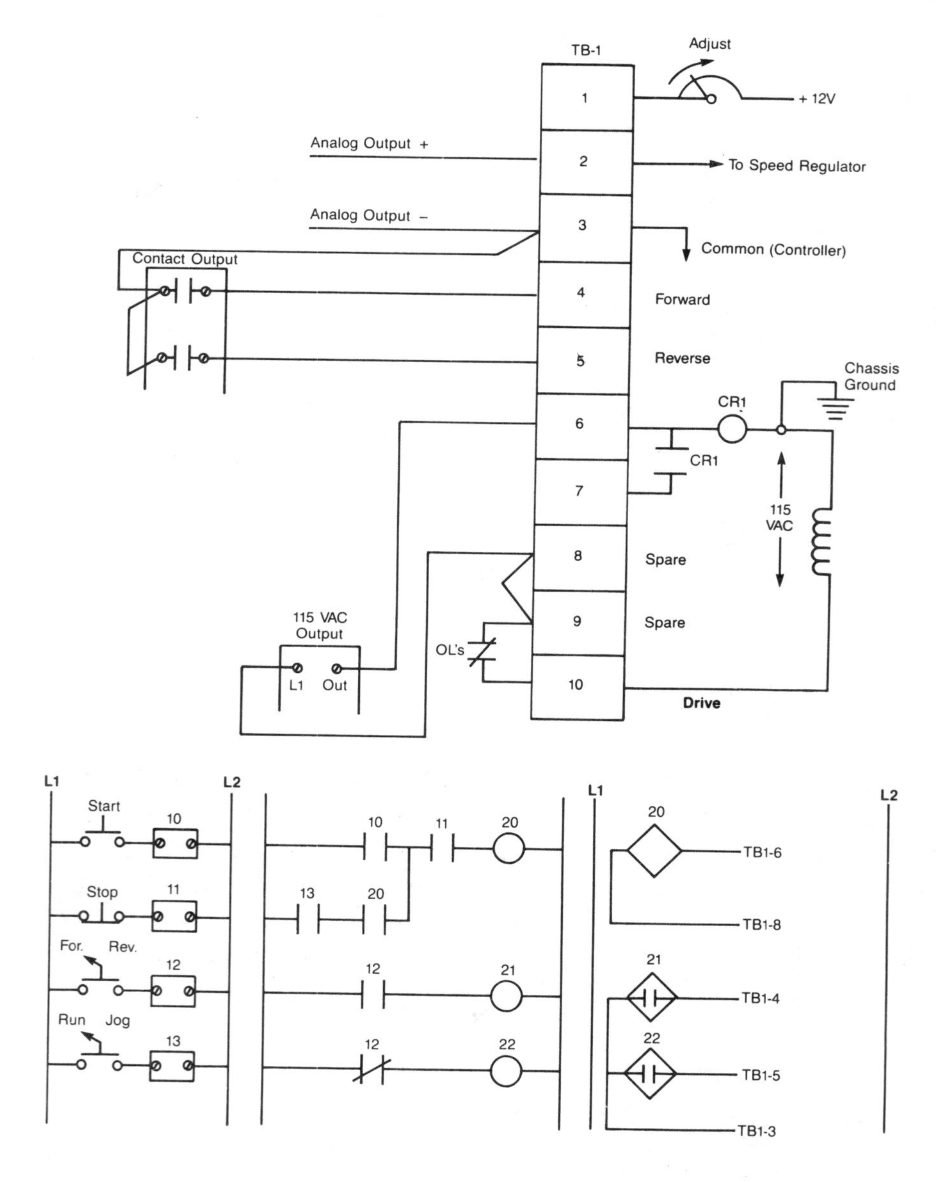

8-43 There are several methods of implementing an AUTO/MANUAL control utilizing the existing operator's manual station. The first method uses a two position selector switch (AUTO/MANUAL) for switching control between the PC and the operator's station. This first configuration is similar to the solution of problem 8-42; however, if the PC is required to have an AUTO/MANUAL indication, one of the two position selector switches must be connected to an input module in addition to the START/STOP, RUN/JOG, and FORWARD/REVERSE signals.

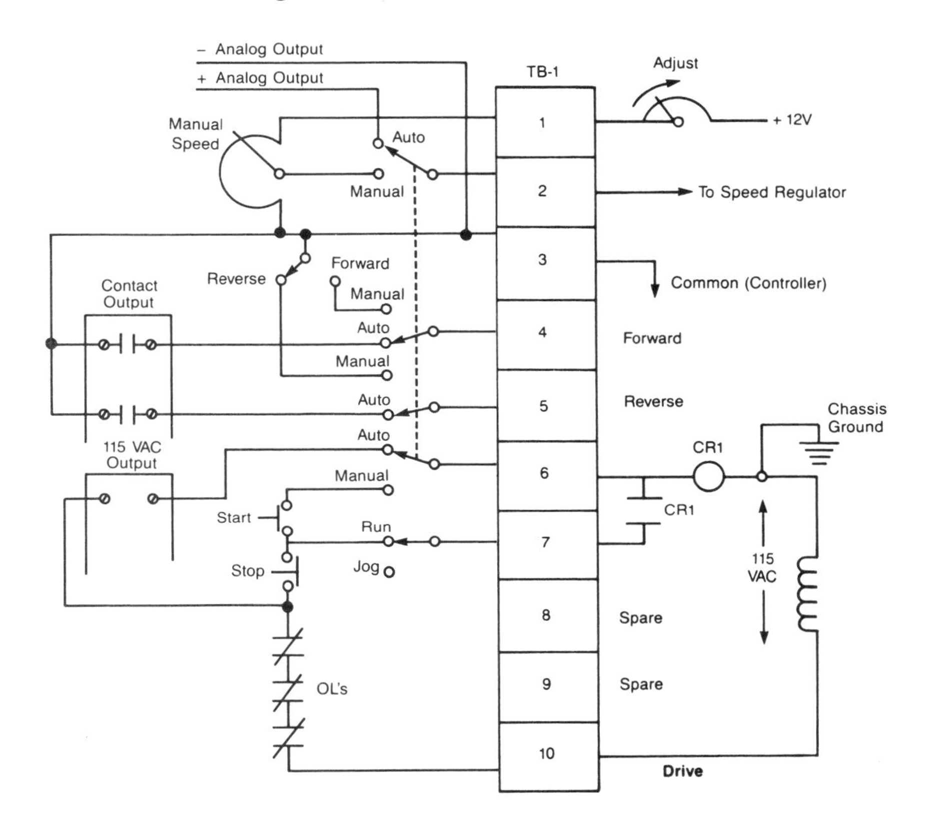

A second configuration, alternate solution, does away with the use of the AUTO/MANUAL connections except for one, the one connected to the PC input module. In this configuration, when the command is in AUTO, the outputs for the AUTO switching signals will be turned ON so that the contacts close and the signals are controlled by the PC program. If contact output modules can be selected with NO or NC contacts, the manual signals should use the NC. This configuration, however, is not completely manual since the PC still has to be operating (RUN) to switch even the manual control.

Alternate solution:

8-43 Continued

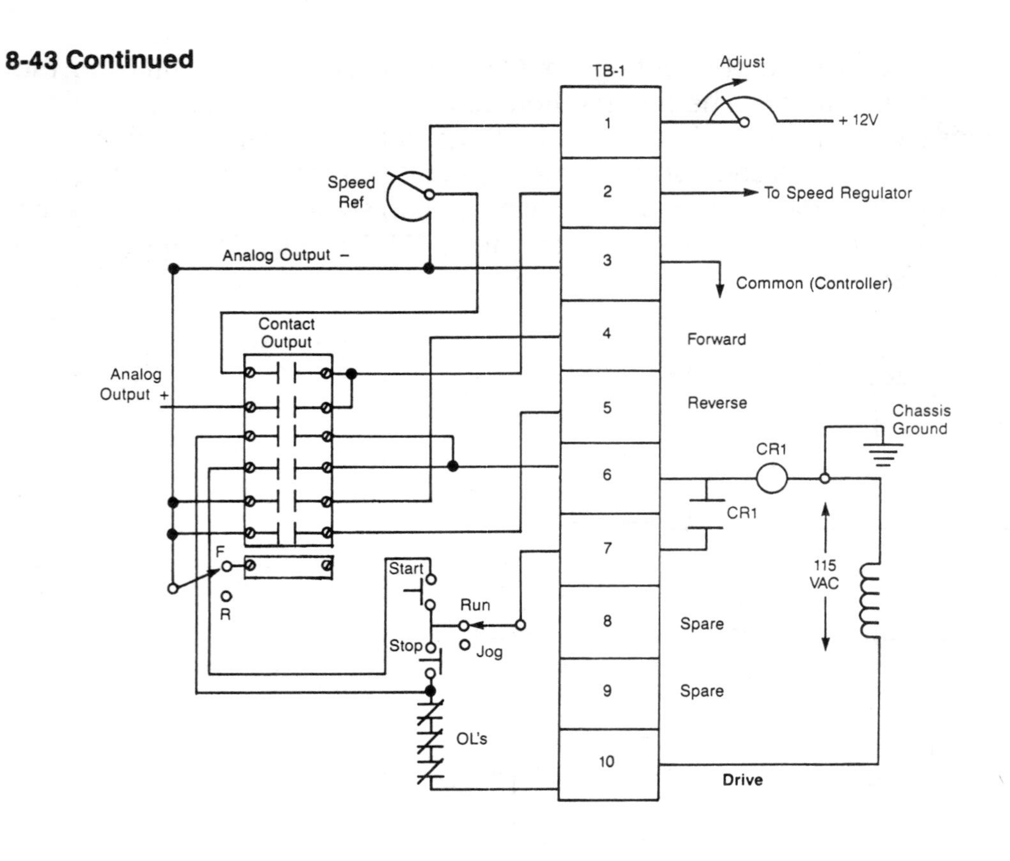

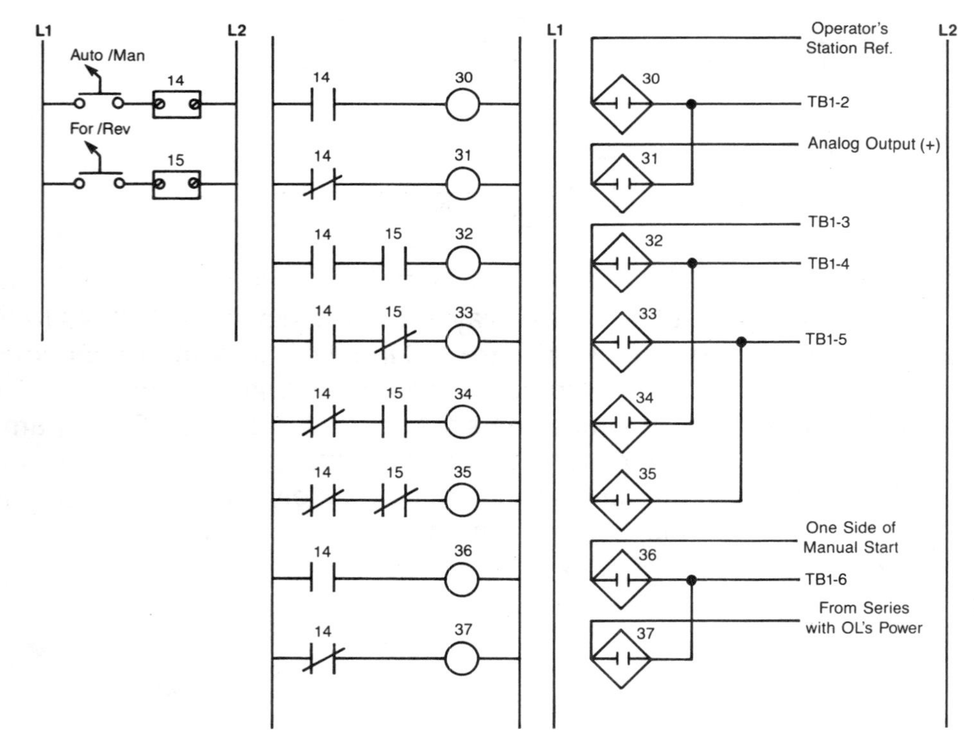

8-44

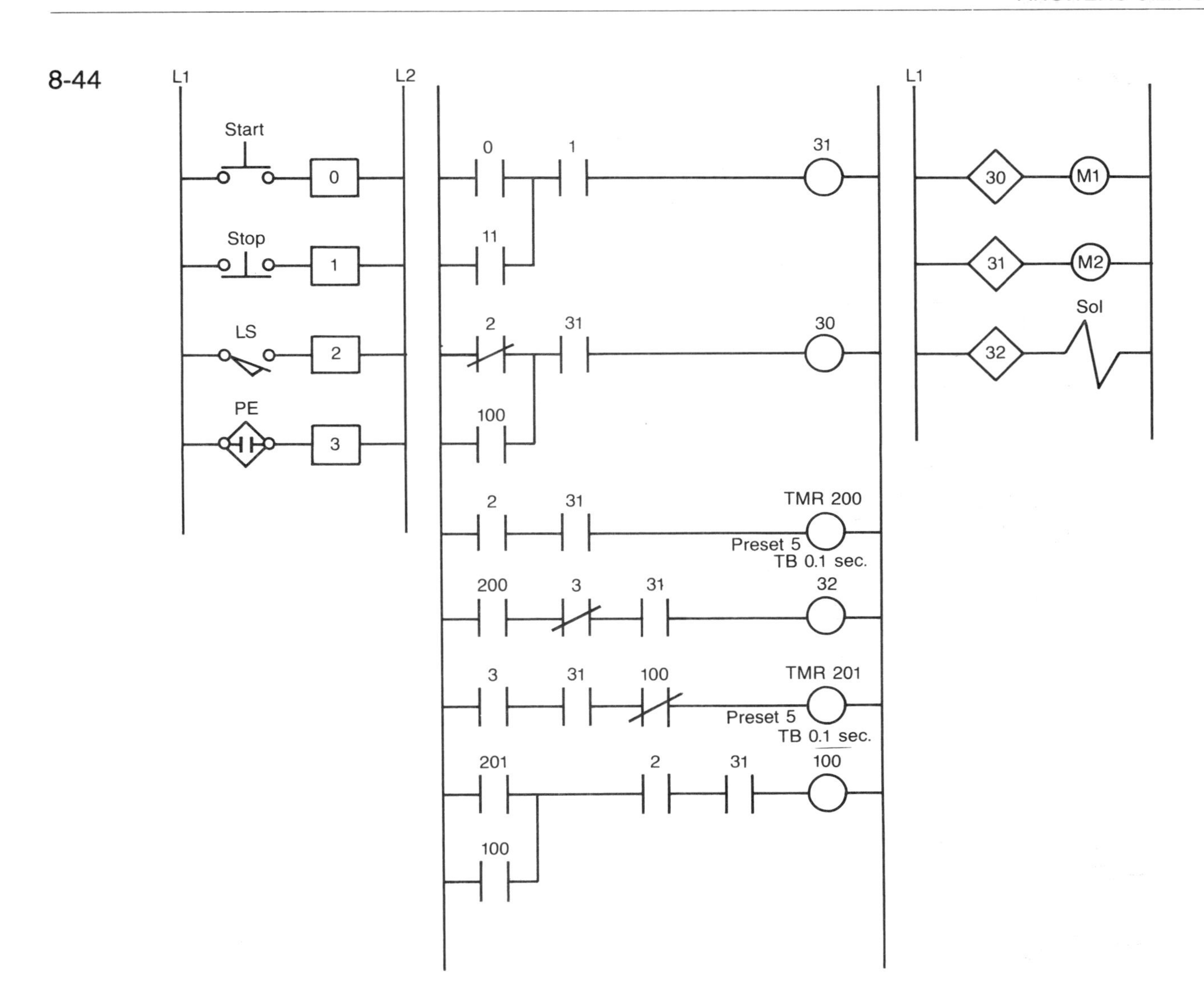

8-45 In the PC ladder program, note that CR4 contact NC in line 11 has bidirectional effect that must be separated. The logic rungs in line 10 and 12 interact inside and outside of the MCR control. Therefore, internal 1003 is set up inside the MCR zone for this purpose. Solenoid 1 (output 32) can be energized inside or outside the MCR zone. Internal 1003 is also used for the SOL1 control.

8-45 Continued

Portions to leave hardwired:

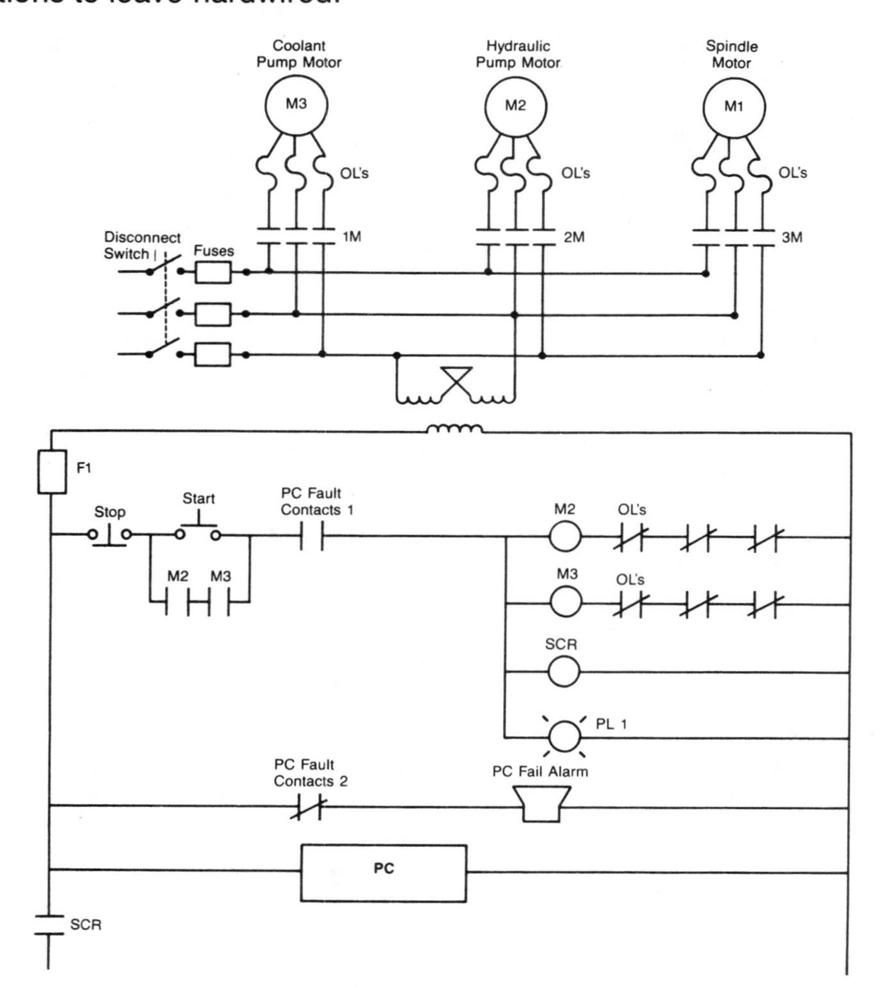

Internal Output Address Assignment

Device	Internal	Description
CR1	—	Same as M1 Address (30)
CR2	1000	Replace CR2
MCR	MCR2000	First MCR Address - Replace MCR
CR3	1001	Replace CR3
CR4	—	Same as PL4 Address (35)
CR5	—	Same as Sol3 Address (36)
—	1002	Trap Timer Logic for Instant Contact
TDR1	T 2040	First Timer Address - Replace TDR1
MCR⊣ ⊢	END 2000	END MCR Logic Section
—	1003	Internal for Inside MCR Logic for Sol1.

8-45 Continued

Real I/O Address Assignment

Module Type	I/O Address Rack	I/O Address Group	I/O Address Terminal	Description
Input	0	0	0	Set-Up/Run
	0	0	1	Run PB
	0	0	2	Up LS1
	0	0	3	Enable SS
Input	0	0	4	Up PB
	0	0	5	LS2
	0	0	6	LS3
	0	0	7	Feed LS4
Input	0	1	0	LS5
	0	1	1	Not Used
	0	1	2	Not Used
	0	1	3	Not Used
	0	1	4	Spare
	0	1	5	
	0	1	6	
	0	1	7	
	0	2	0	Spare
	0	2	1	
	0	2	2	
	0	2	3	
	0	2	4	Spare
	0	2	5	
	0	2	6	
	0	2	7	
Output	0	3	0	M1 Starter
	0	3	1	PL2 Master On
	0	3	2	Sol1 Up
	0	3	3	Sol2 Down
Output	0	3	4	PL3 Down On
	0	3	5	PL4 Set Up On
	0	3	6	Sol3 Feed
	0	3	7	Sol4 Fast Feed
	0	4	0	Spare
	0	4	1	
	0	4	2	
	0	4	3	

8-45 Continued

Ladder Program

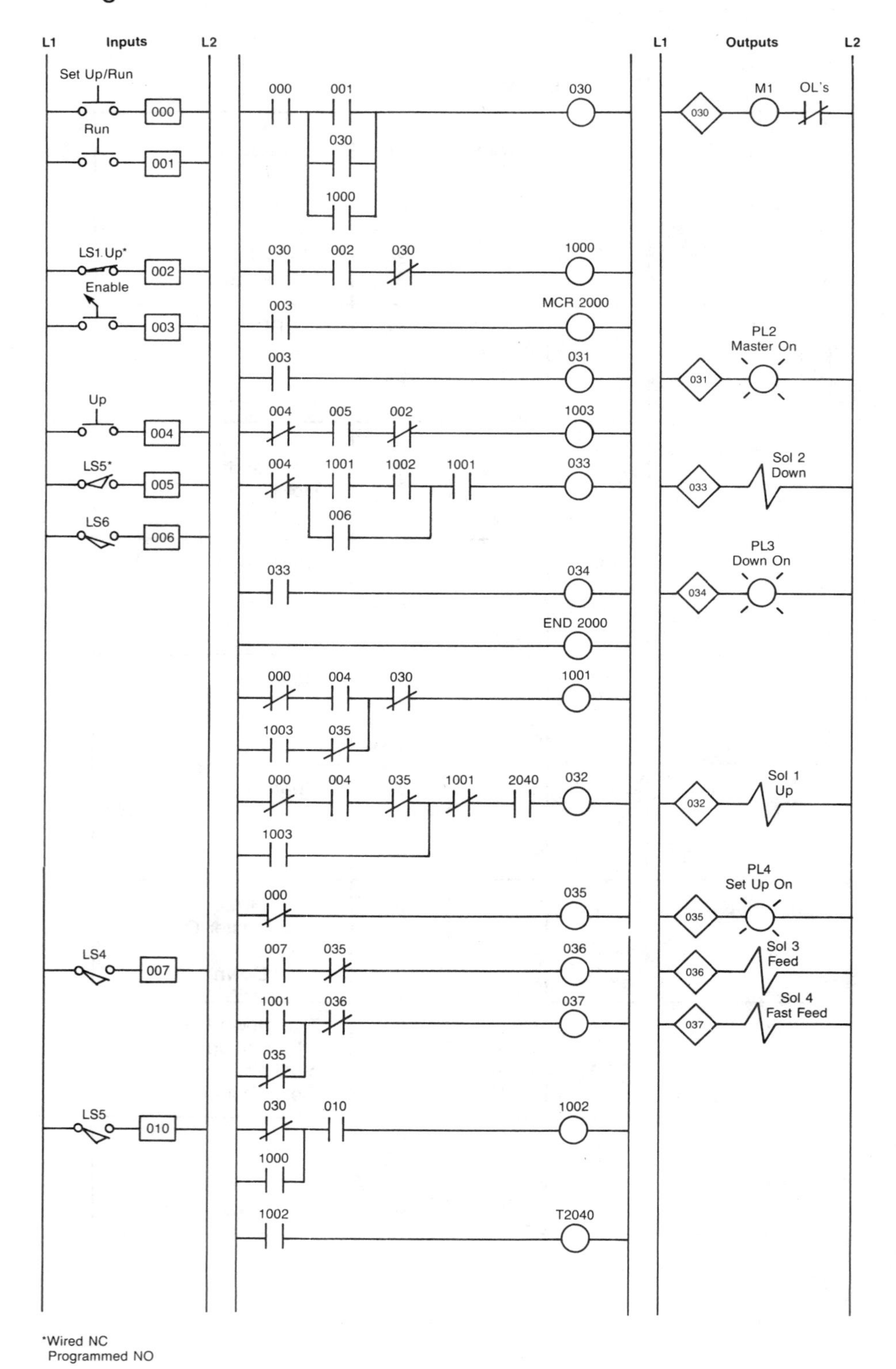

*Wired NC
Programmed NO

8-46

a) Flowchart for Process

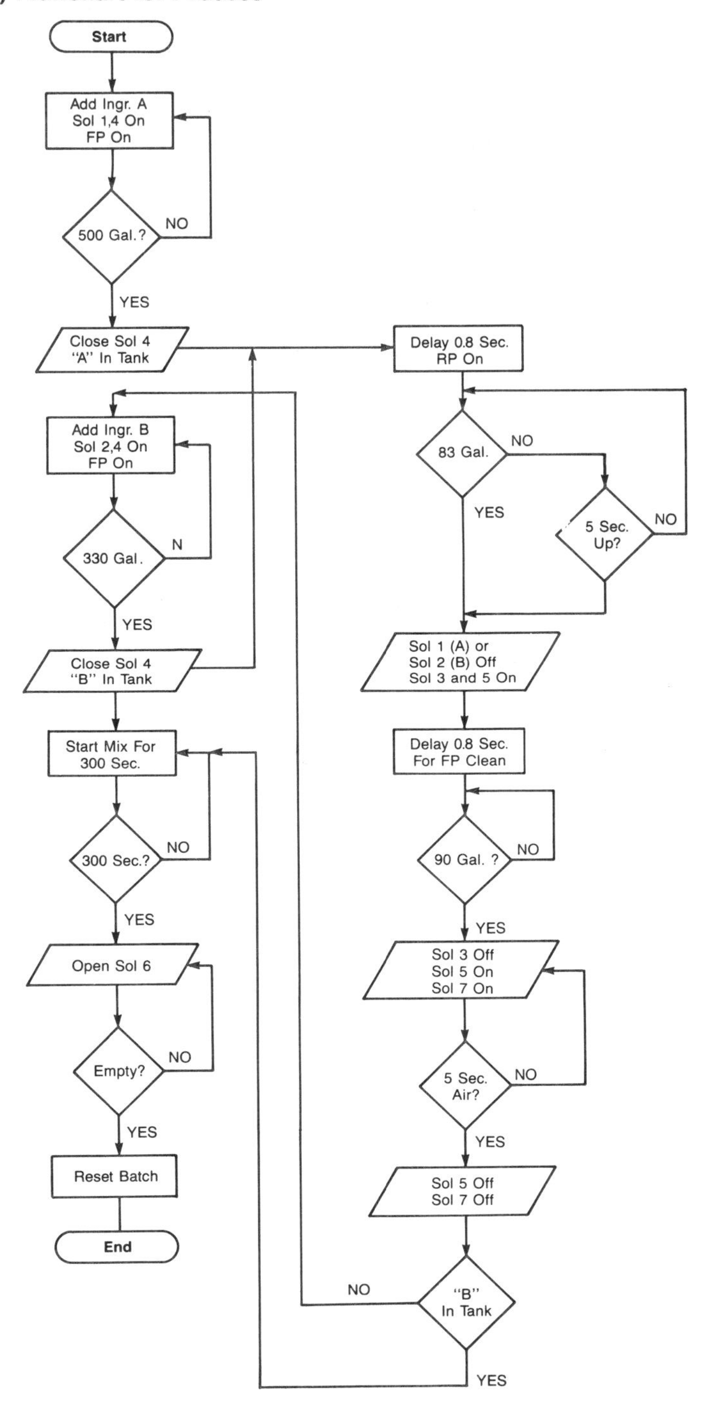

8-46 Continued

b) Logic Implementation of Process:

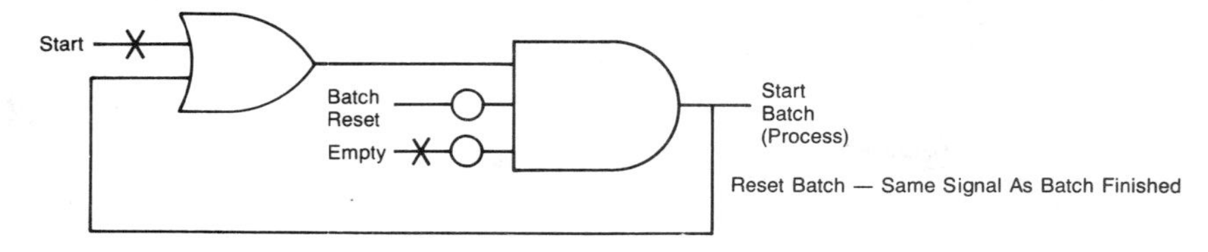

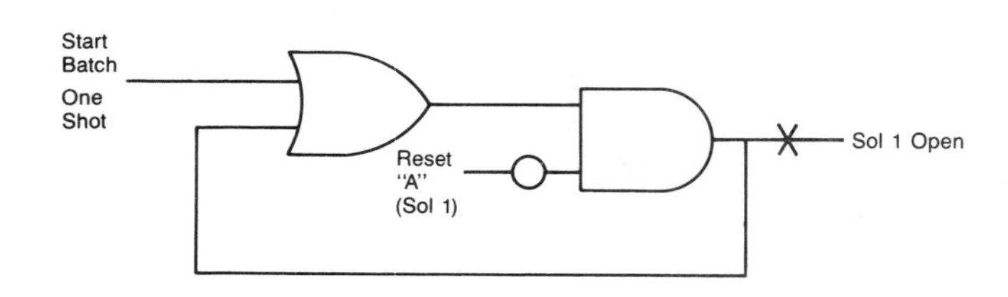

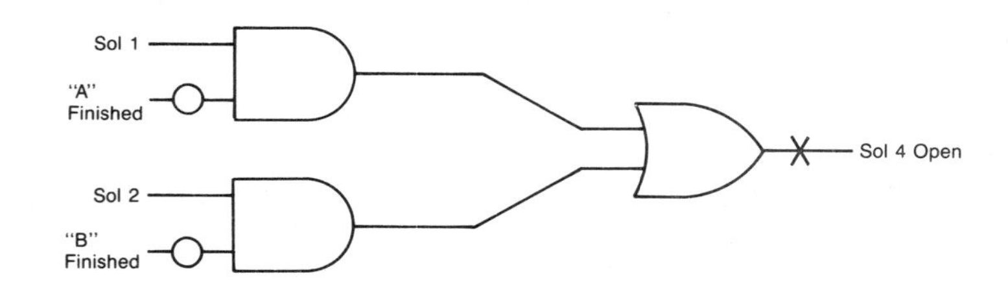

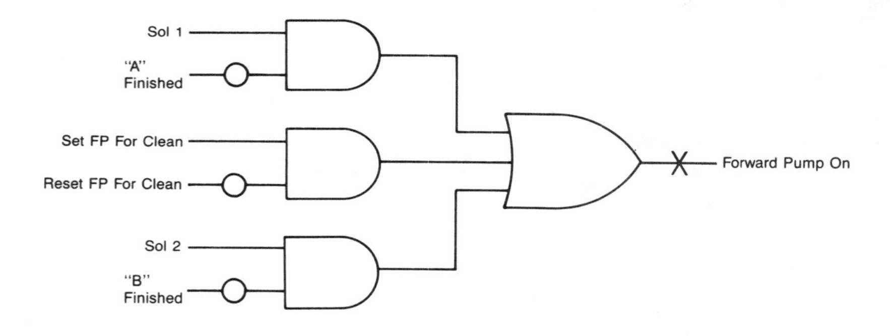

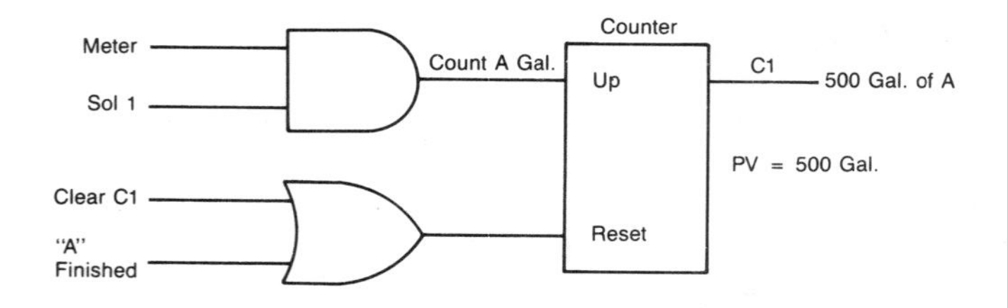

8-46 b) Continued

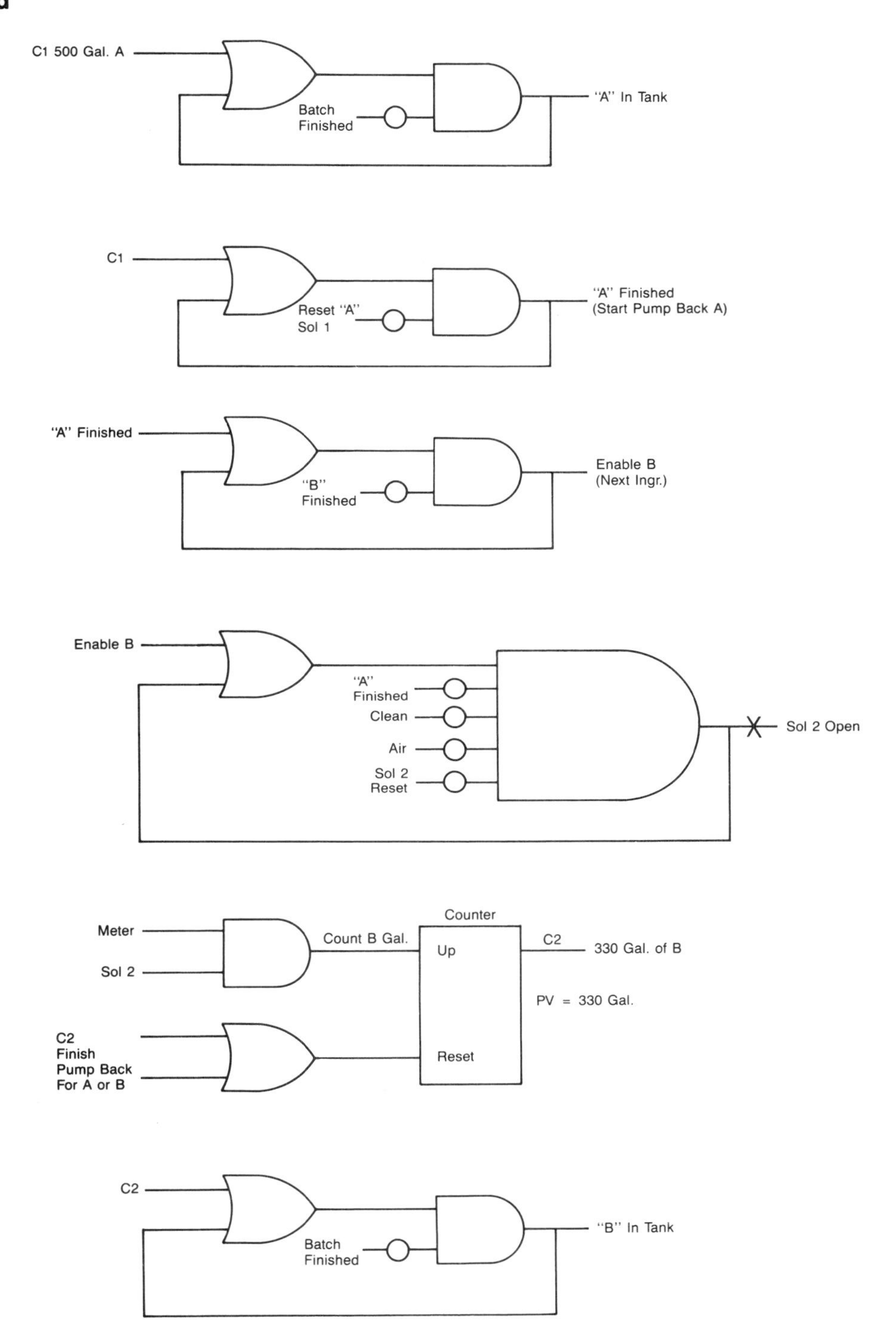

8-46 b) Continued

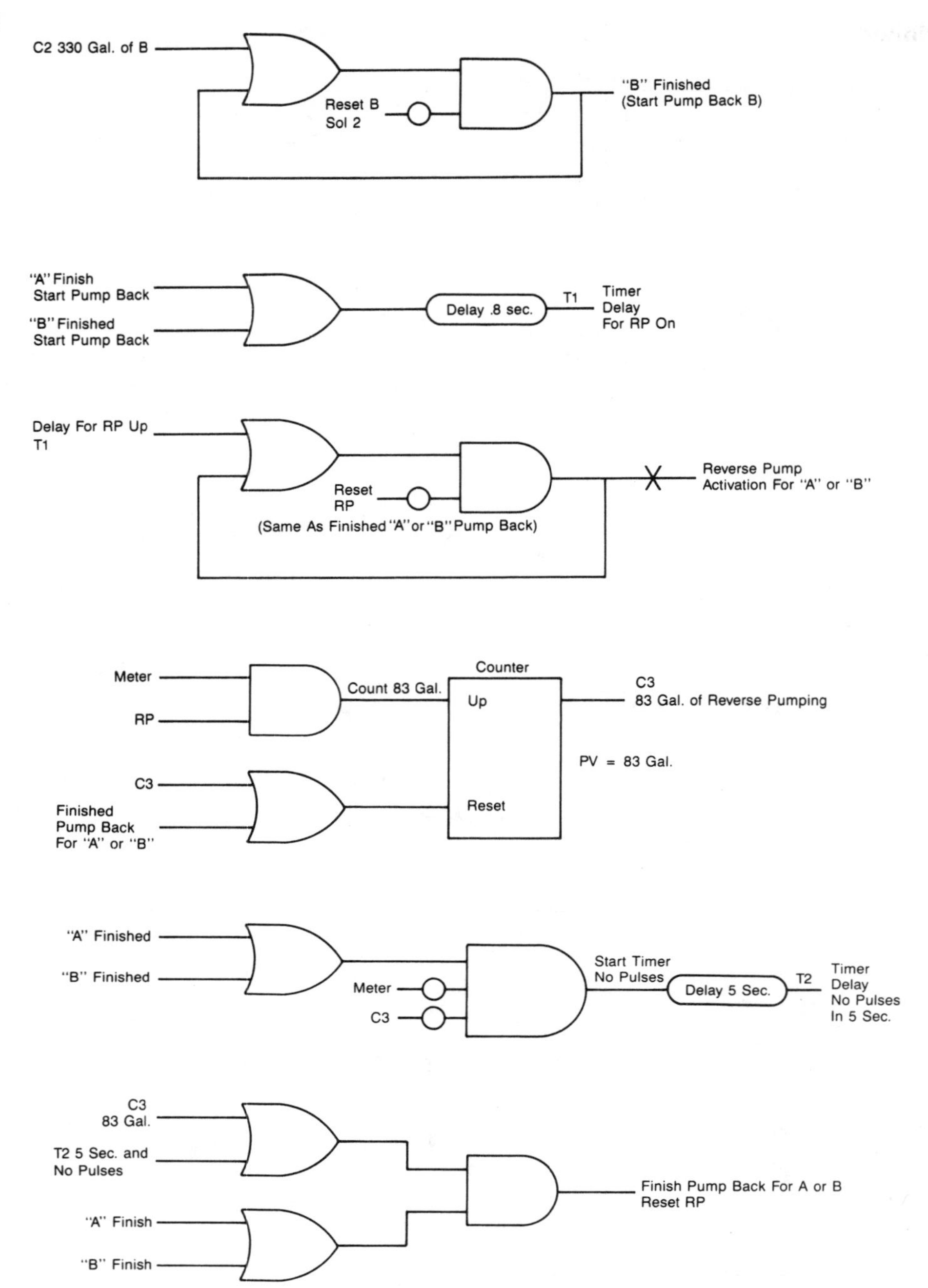

8-46 b) Continued

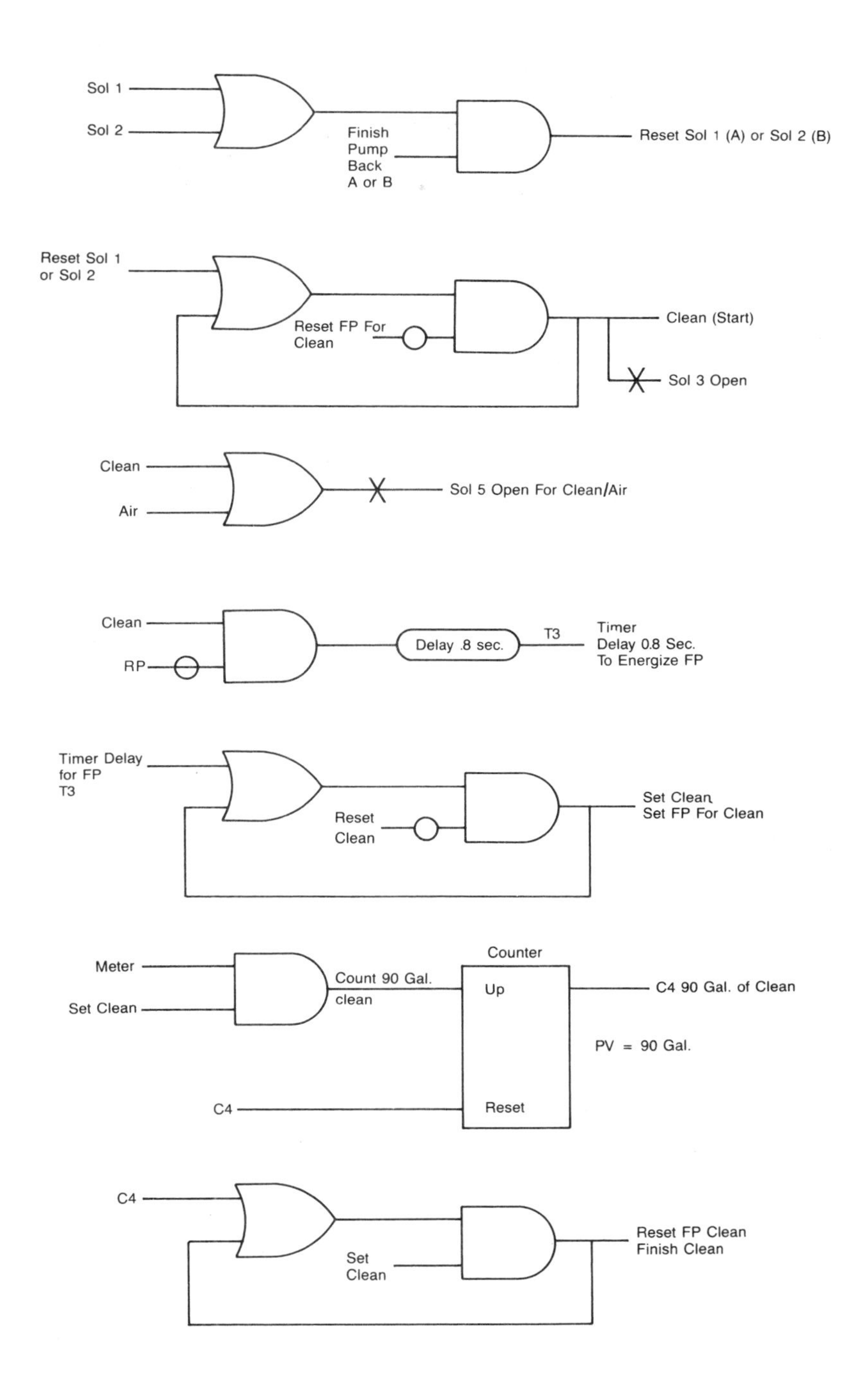

8-46 b) Continued

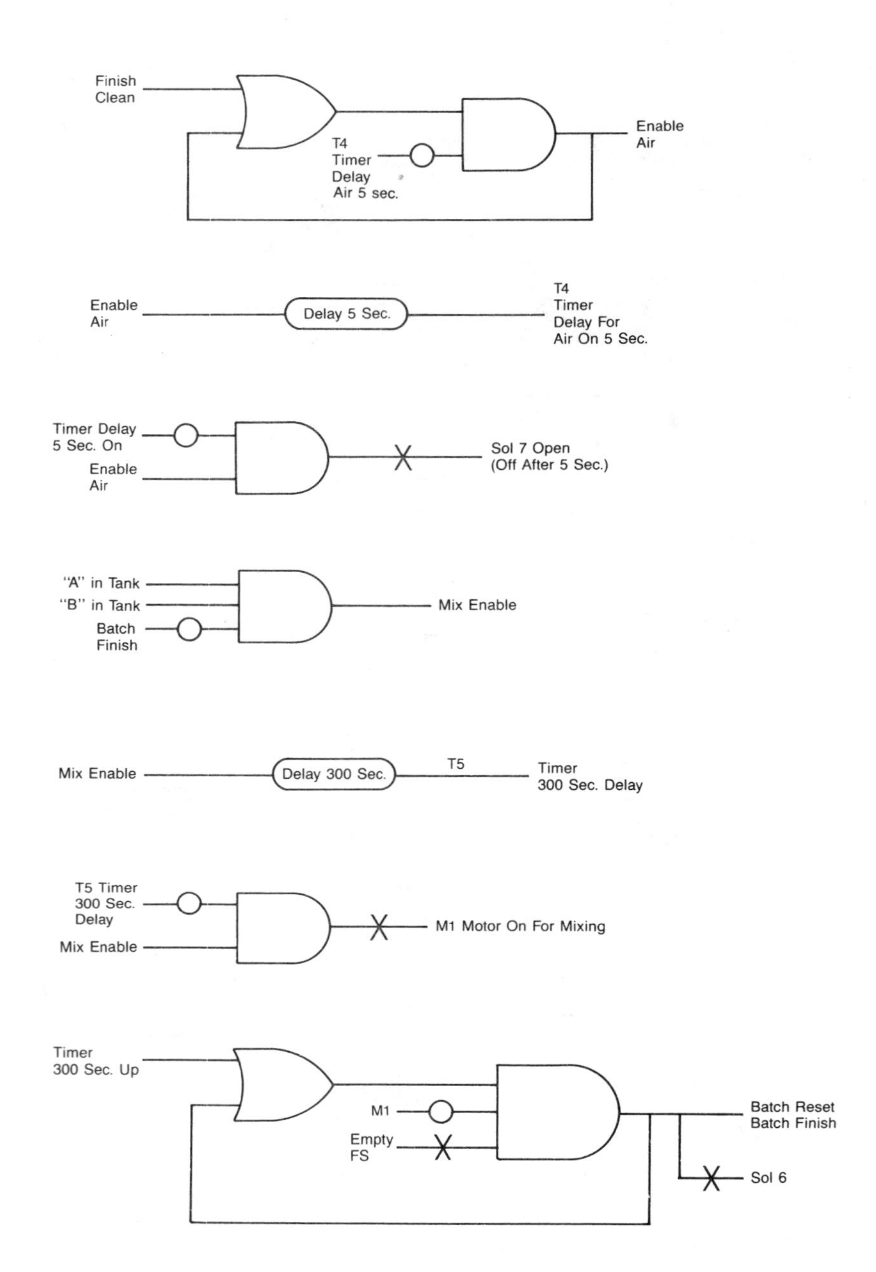

8-46

c) Address Assignment:

Module Type	I/O Address Assignment Rack	Group	Terminal	Description
Output	0	0	0	Sol 1 Ingr. A
	0	0	1	Sol 2 Ingr. B
	0	0	2	Sol 3 Cleanser
	0	0	3	Sol 4 Valve
Output	0	0	4	Sol 5 Valve
	0	0	5	Sol 6 Valve
	0	0	6	Sol 7 Air
	0	0	7	M1 Mix Motor
Output	0	1	0	Forward Pump
	0	1	1	Reverse Pump
	0	1	2	
	0	1	3	
	0	1	4	Spare
	0	1	5	
	0	1	6	
	0	1	7	
Input	0	2	0	Start Batch PB Ops.
	0	2	1	Float Switch FS1
	0	2	2	Meter Pulses
	0	2	3	
	0	2	4	Spare
	0	2	5	
	0	2	6	
	0	2	7	
	0	3	0	Spare
	0	3	1	
	0	3	2	
	0	3	3	
	0	3	4	Spare
	0	3	5	
	0	3	6	
	0	3	7	
	0	4	0	Spare
	0	4	1	
	0	4	2	
	0	4	3	

Internal Assignment

DEVICE	INTERNAL	DESCRIPTION
Logic	1000	Start Batch
Logic	1001	Start Batch One-shot
	—	Reset Batch — Batch Finished Same as Sol 6 (Output 005)
	1002	Reset "A", Sol 1; Reset "B", Sol 2
	1003	"A" Finished — Start Pump Back A
	1004	"B" Finished — Start Pump Back B
	1005	Set FP for Clean
	1006	Reset FP for Clean — Finish Clean — Reset
Counter	C2300	Counter for Ingr A (500 Gal. Preset) C1
	1007	"A" In Tank
	1010	Enable B (Next Ingr.)
	1011	Clean (Start)
	1012	Air — Enable Air
	1013	Sol 2 Reset
	C2301	Counter for Ingr. B (330 Gal. Preset) C2
	1014	"B" In Tank
	T2000	Timer T1, 0.8 Sec. Delay for RP
	1015	Reset RP, Finish Pump Back for A or B
	C2302	Counter for Reverse Pump, 83 Gal., C3
	T2001	Timer T2, 5 Sec. Preset — No Pulses In 5 Sec.
	T2002	Timer T3, 0.85 Sec. Delay for FP
	C2303	Counter for Clean — 90 Gal. — C4
	T2003	Timer T4, 5 Sec . Air On
	1016	Mix Enable
	T2004	Timer T5, 300 Sec. Motor On

8-46

d) PC Ladder Diagram and I/O Connection diagram:

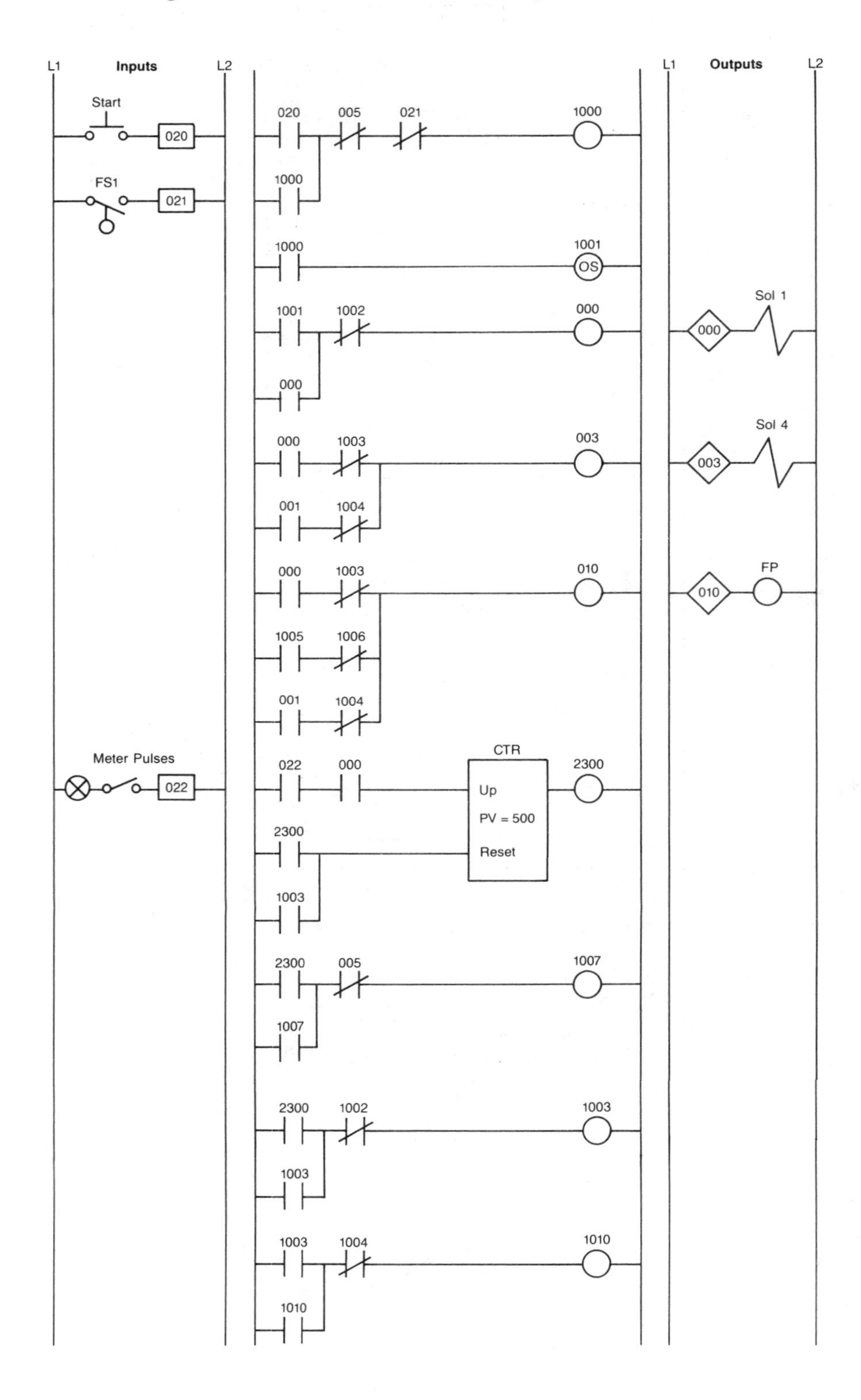

8-46 d) Continued

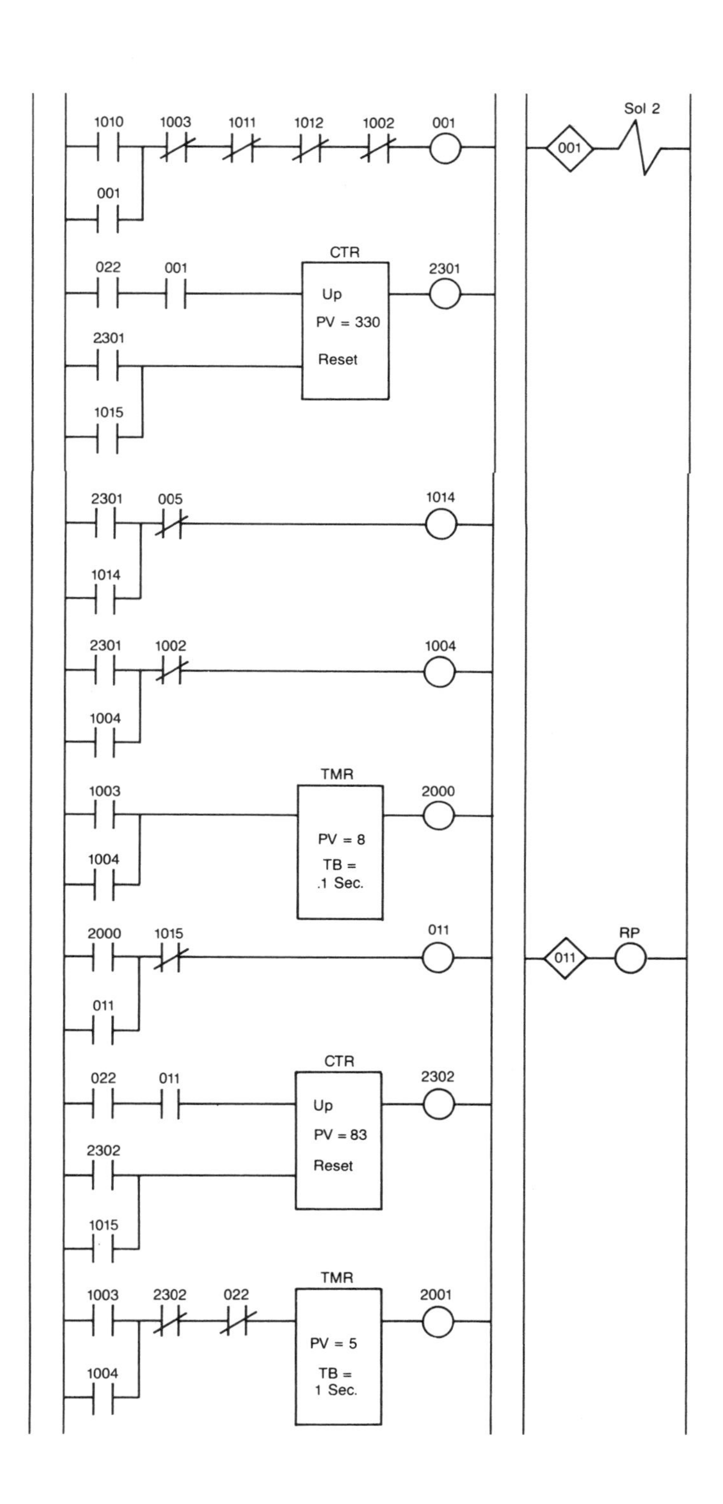

8-46 d) Continued

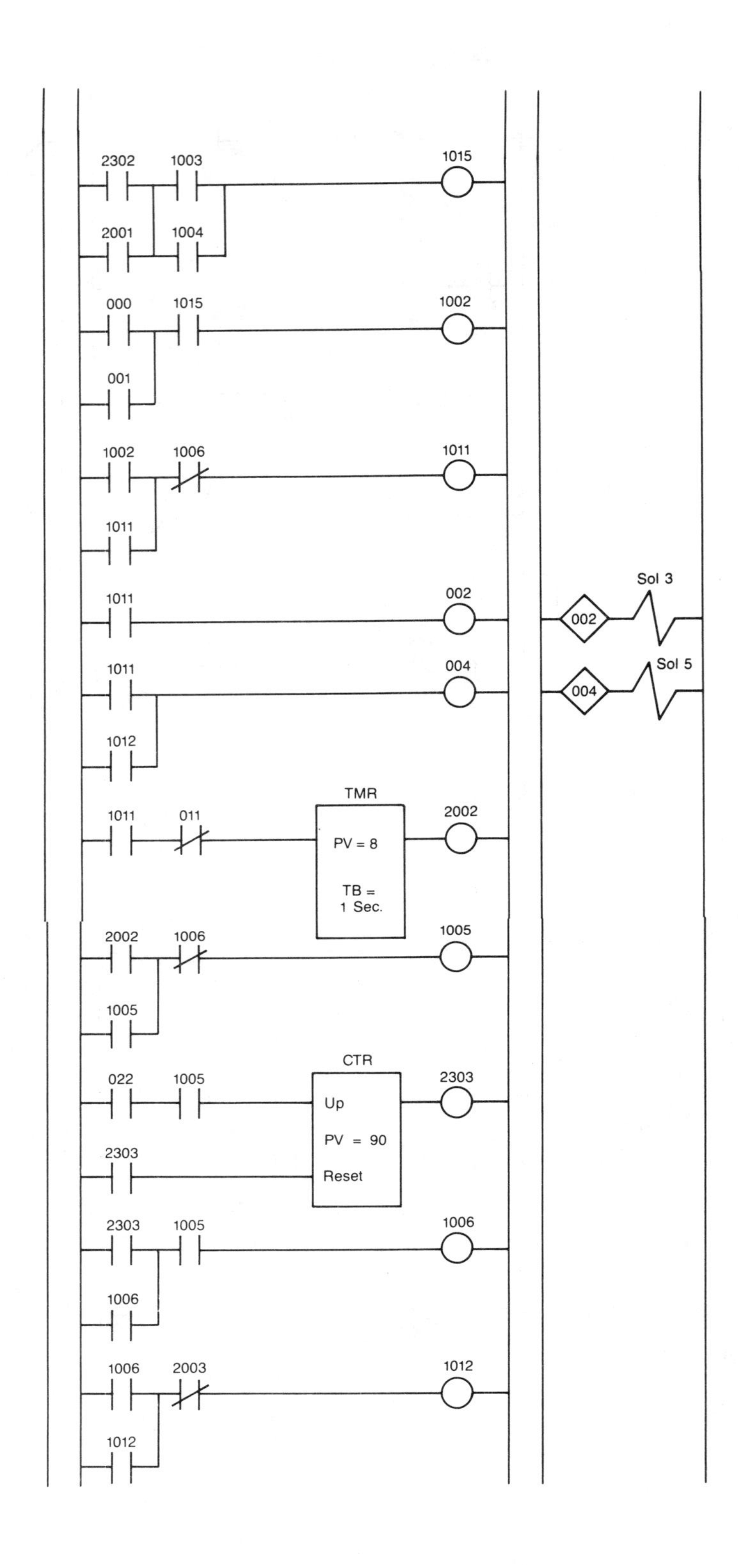

8-46 d) Continued

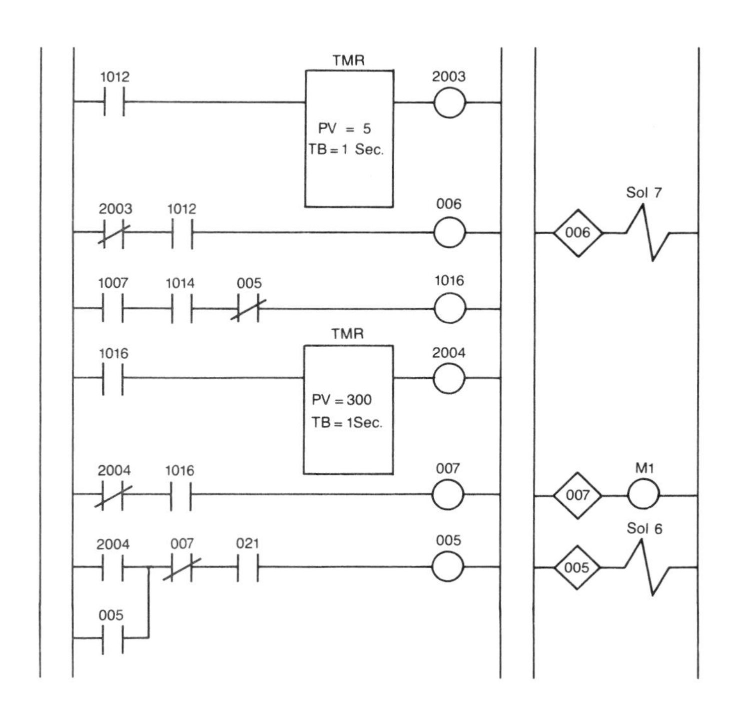

8-47 The Exclusive - OR is expressed as $Z = \overline{A} \bullet B + A \bullet \overline{B}$

Using Boolean algebra we get:

$$Z = \overline{A \cdot \overline{B} + \overline{A} \cdot B}$$

$$= \overline{A \cdot \overline{B}} + \overline{\overline{A} \cdot B}$$

$$= \overline{A} + \overline{\overline{B}} \cdot \overline{\overline{A}} \cdot \overline{B}$$

$$= (\overline{A} + B) \cdot (A + \overline{B})$$

The ladder implementation is:

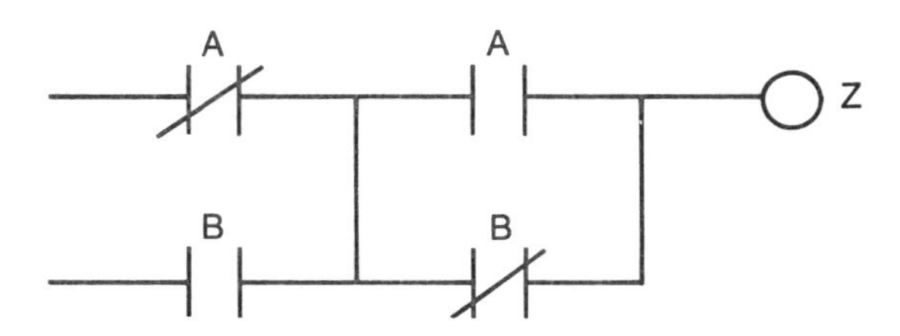

Truth Table

A	B	Z
0	0	1
0	1	0
1	0	0
1	1	1

8-48 True. If a transitional or one-shot output is energized at the middle of a scan, it will be ON for half a scan.

8-49 True. The one-shot pulse will never be seen since in the evaluation of the internal 1 in the next scan the contacts of the one-shot will be OFF. The one-shot is ON for one or fewer scan times from the moment it is energized.

8-50

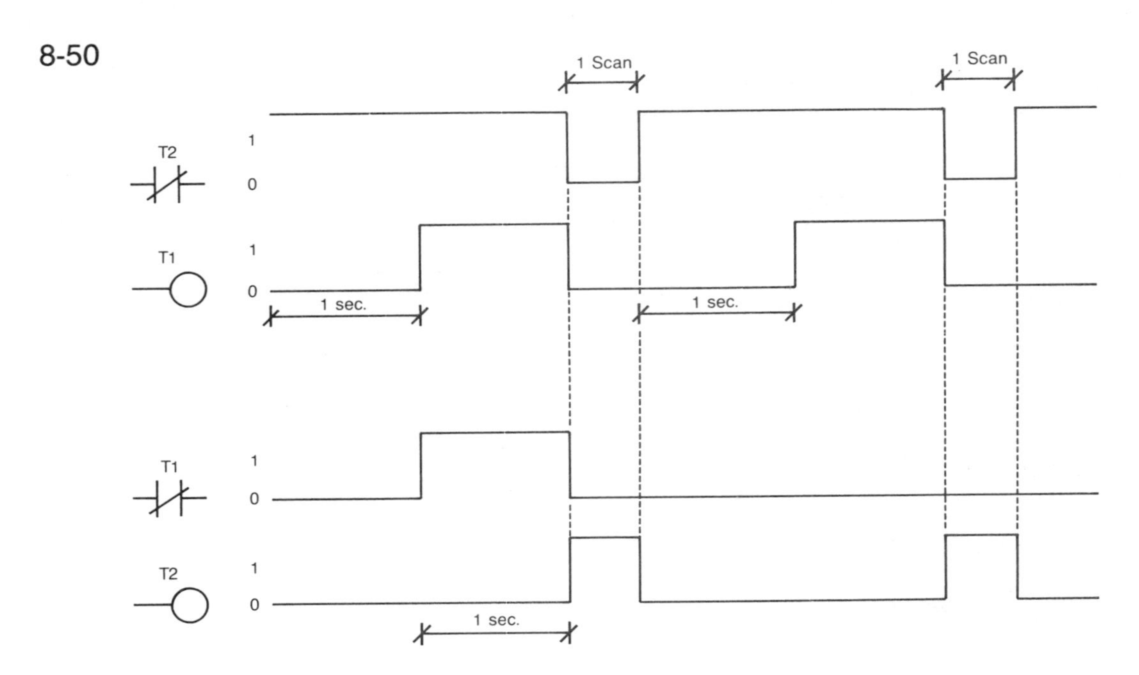

8-51 True. The resetting of T1 would occur in the next scan after T2 is energized.

8-52 The output of T1 is ON for one scan and is used to reset T2. It is not possible to see it on a CRT since the time ON of T1 is very short.

8-53

T2 — T1

Preset Value = 1
Time Base = 1 sec.

T1 — T2

Preset Value = 1
Time Base = 1 sec.

T2 — Reset — TRAP

TRAP

8-54

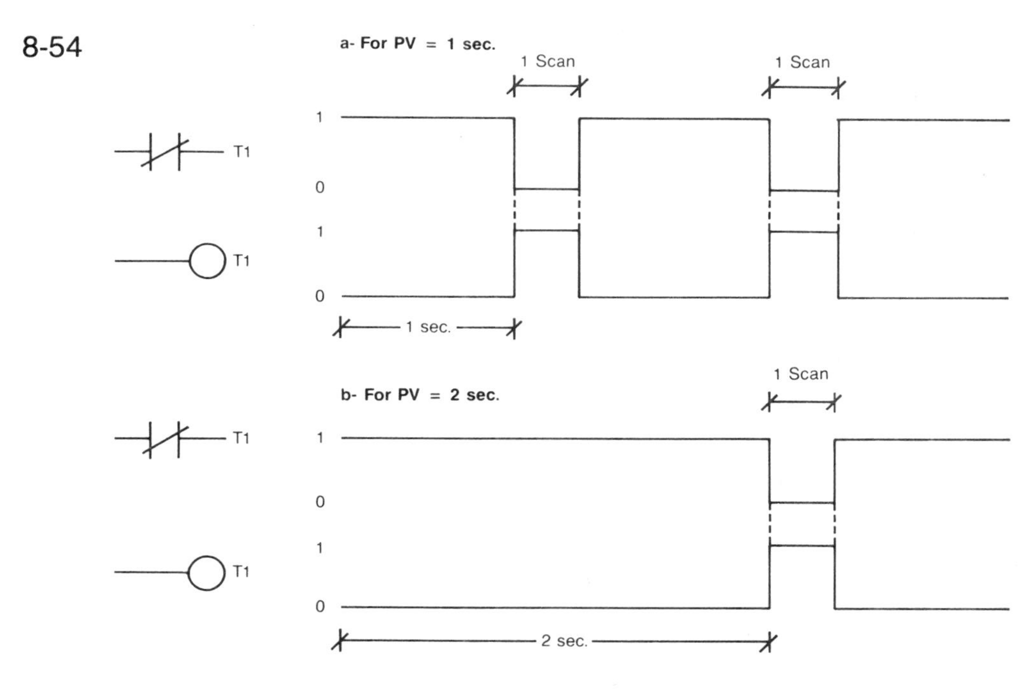

8-55 c- PV = 96 for accuracy after first 1 sec cycle

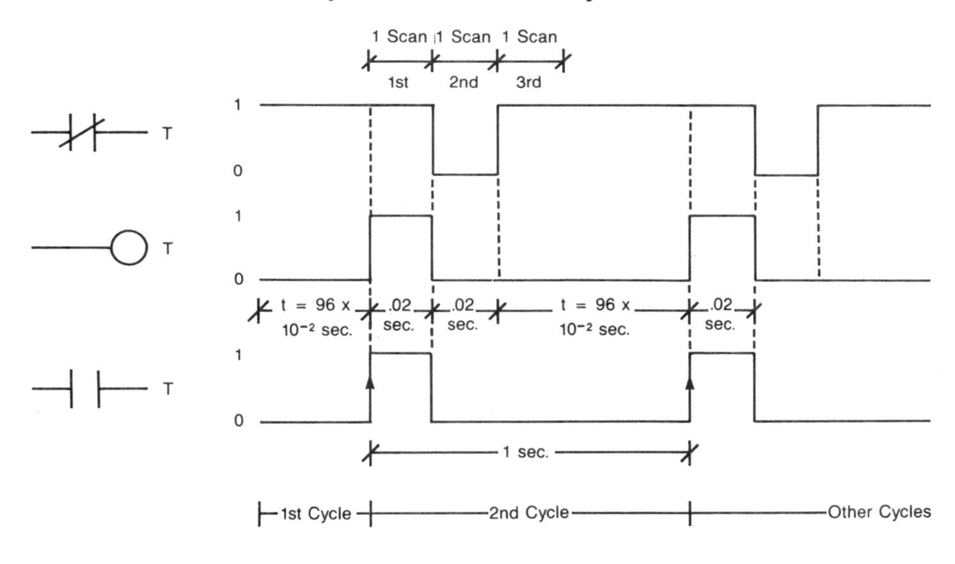

Contacts of time used to trigger counter inputs

8-56

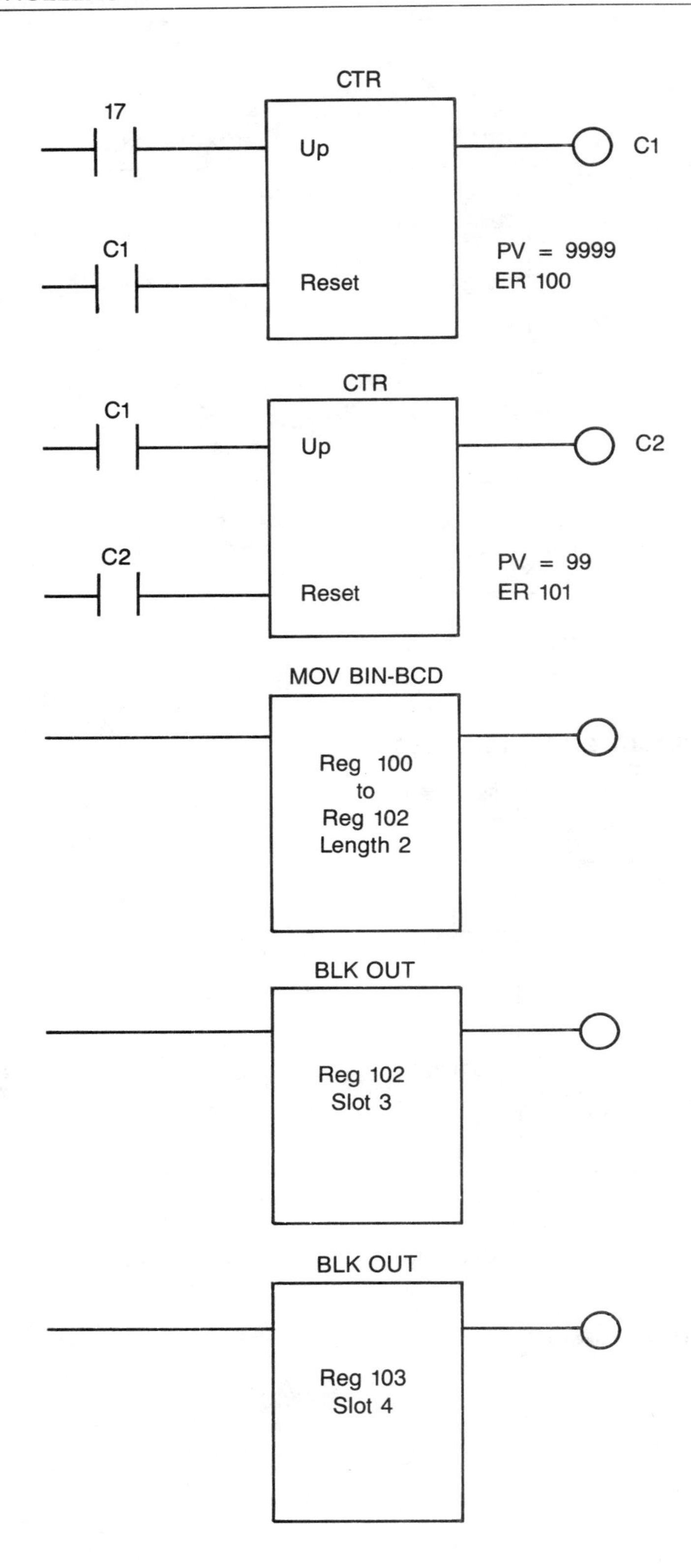
CTR
17
Up
C1
C1
Reset
PV = 9999
ER 100
CTR
C1
Up
C2
C2
Reset
PV = 99
ER 101
MOV BIN-BCD
Reg 100
to
Reg 102
Length 2
BLK OUT
Reg 102
Slot 3
BLK OUT
Reg 103
Slot 4

8-57

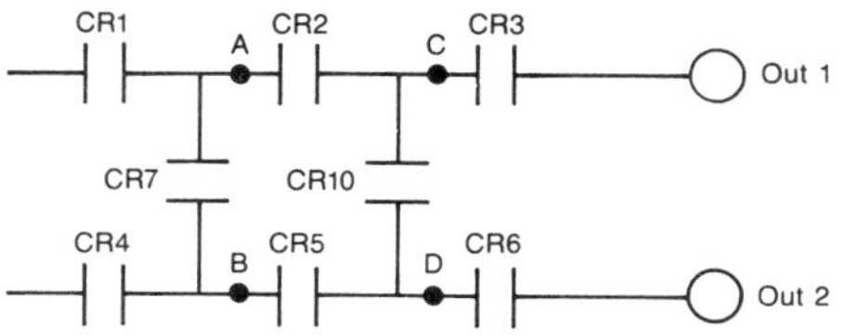

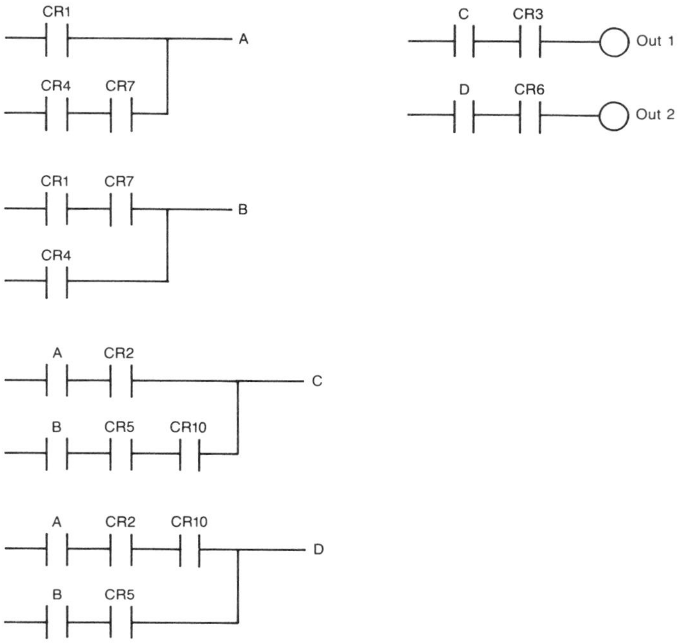

Replacing each group of letter contacts by its logic we get:

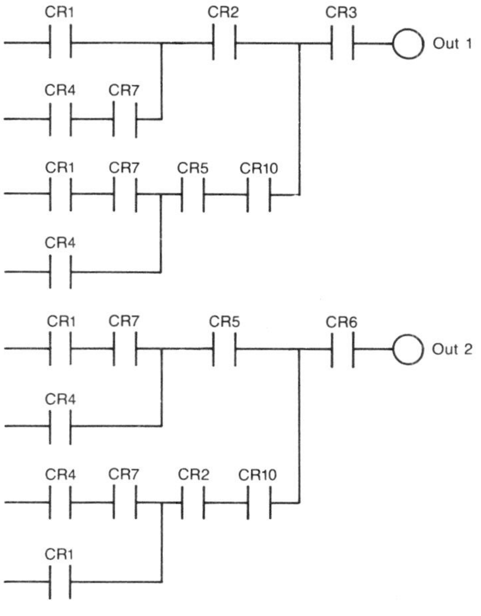

8-58

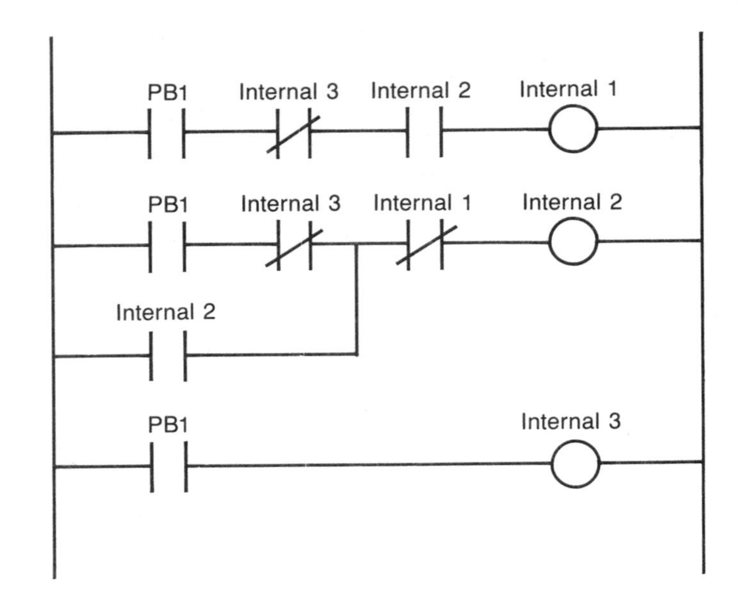
PB1
Internal 3
Internal 2
Internal 1
PB1
Internal 3
Internal 1
Internal 2
Internal 2
PB1
Internal 3

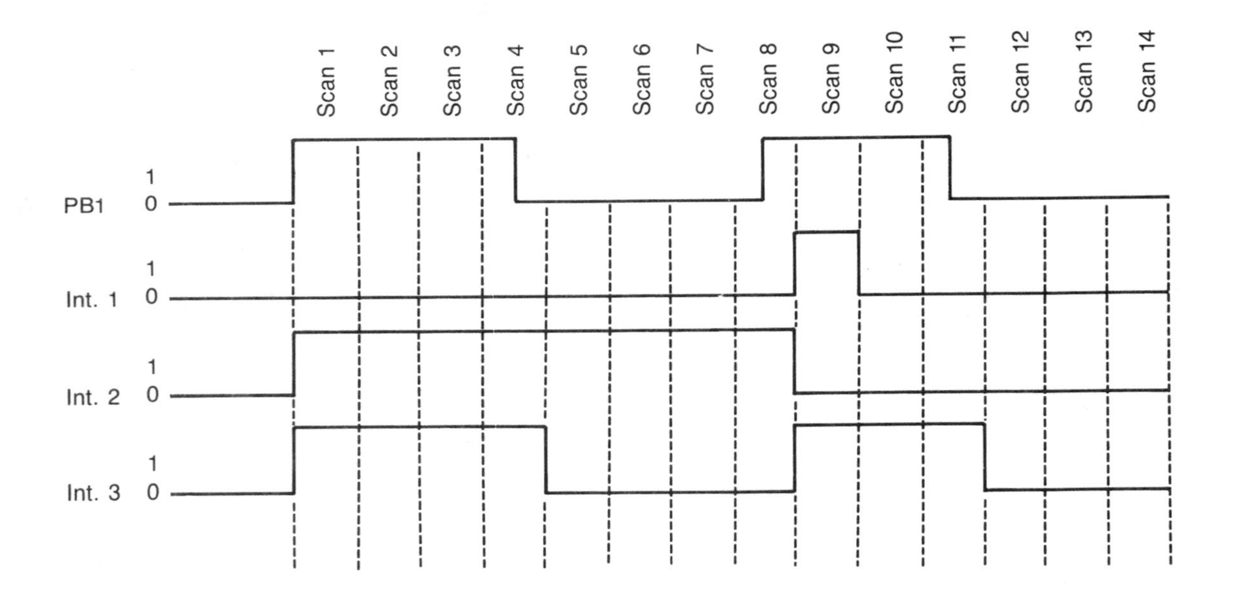
Scan 1
Scan 2
Scan 3
Scan 4
Scan 5
Scan 6
Scan 7
Scan 8
Scan 9
Scan 10
Scan 11
Scan 12
Scan 13
Scan 14
PB1
Int. 1
Int. 2
Int. 3
1
0

8-59 f- b and c. Internal detects the second push of PB1 and is ON for one scan in a one-shot fashion.

8-60 False. The circuit will not work. If the logic of internal 1 is changed, it will not be able to reset using the one-shot. If the logic of internal 2 is changed, it will not detect the first push of PB1.

UNIT

--]9[--

ANSWERS

9-1 Documentation

9-2 False. The documentation process starts from the moment a project starts.

9-3 False. Software as well as hardware information is used in a documentation package.

9-4 d- Usage of ladder instructions

9-5

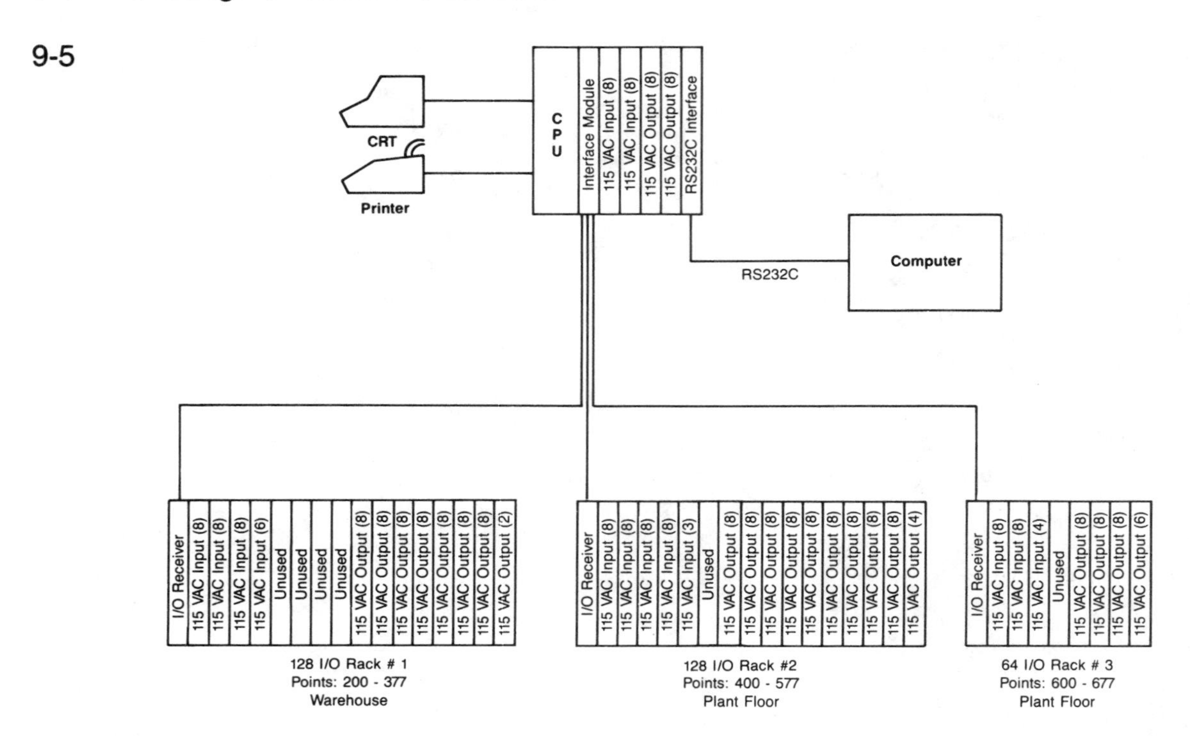

9-6 I/O address assignment

	I/O Address			
Module Type	**Rack**	**Group**	**Terminal**	**Description**
Input 115 VAC	1	3	0	PB10
	1	3	1	LS2
	1	3	2	TS7
	1	3	3	LS16 (Wired NC)
	1	3	4	PS4
	1	3	5	Not Used
	1	3	6	Not Used
	1	3	7	Not Used
	⋮	⋮	⋮	⋮
Output 115 VAC	3	2	0	Sol3
	3	2	1	PL4
	3	2	2	PL5
	3	2	3	M4
	3	2	4	Not Used
	3	2	5	Not Used
	3	2	6	Sol7
	3	2	7	Sol10

9-7 b- System abstract

9-8 False. To understand the total control program, one has to have the total document package. It is very difficult, if not impossible, to know what is going on by simply looking at a printout (standard).

9-9 c- I/O wiring connection diagram

9-10 False. The documentation package is useful during system design, installation, start-up, debugging, and maintenance.

9-11 c- a and b

9-12 The address, type of I/O, device, and function or description.

9-13 True

9-14 c- As the internals are used during development

9-15 b- Improper reference use of a register

9-16 c- The I/O connections of field devices

9-17 Cassette tape, floppy disks, and electronic memory modules

9-18 b- Latest ladder printout

9-19 d- All of the above. Documentation systems reduce drafting manpower and time, have extensive program listings, and provide explanations of ladder rungs.

9-20 Program titles
Multiple subtitles
Date and time the documentation was last produced
Page numbering
Extensive commentaries before and after each rung
Contact or element description
PC address for each contact
Pictorial representation of each PC instruction (coils, contacts, etc.)
Rung numbers
Rungs where each contact is used
All preset values of registers used
Identification of internals and real I/O

UNIT
--]10[--

ANSWERS

10-1 The system layout is a conscientious approach in placing and interconnecting the components not only to satisfy the application, but also to insure that the controller will operate trouble free in the environment in which it will be placed.

10-2 False. With proper system layout, the components will be easily accessible.

10-3 True

10-4 Isolation transformers, auxiliary power supplies, safety control relays, circuit breakers, fuse blocks, and line noise suppressors are some of the components that must be taken into consideration during the system layout.

10-5 c- Close to the machine or process

10-6 d- All of the above

10-7 b- Vibration

10-8 The enclosure should be placed in a position that allows the doors to be opened fully for easy access to wiring and components for testing or trouble-shooting.

The enclosure depth should be enough to allow clearance between a closed enclosure door and any print pocket mounted on the door or the enclosed components and related cables.

The enclosure's back panel should be removable to facilitate mounting of the components and other assemblies.

An emergency disconnect device should be mounted in the cabinet in an easily accessible location.

Accessories, such as AC power outlets, interior lighting, or a gasketed plexiglass window to allow viewing of the processor or I/O indicators should be considered for installation and maintenance convenience.

10-9 False. A power outlet is very convenient inside the enclosure since a programming terminal could be plugged in during start-up.

10-10 c- 60°C inside the enclosure. The temperature inside the enclosure will most likely be higher than the temperature outside.

10-11 Fan or blower

10-12 Condensation

10-13 False. It is a good practice to place a PC system far away from high noise generating equipment.

10-14 Vertical position to allow maximum convection cooling.

10-15 Power supply

10-16 d- All of the above

10-17 c- Directly above the CPU

10-18 c- Effects of noise are minimized

10-19 Magnetic starters, contactors, and relays

10-20 False. Fans should be placed in a location near hot spots.

10-21

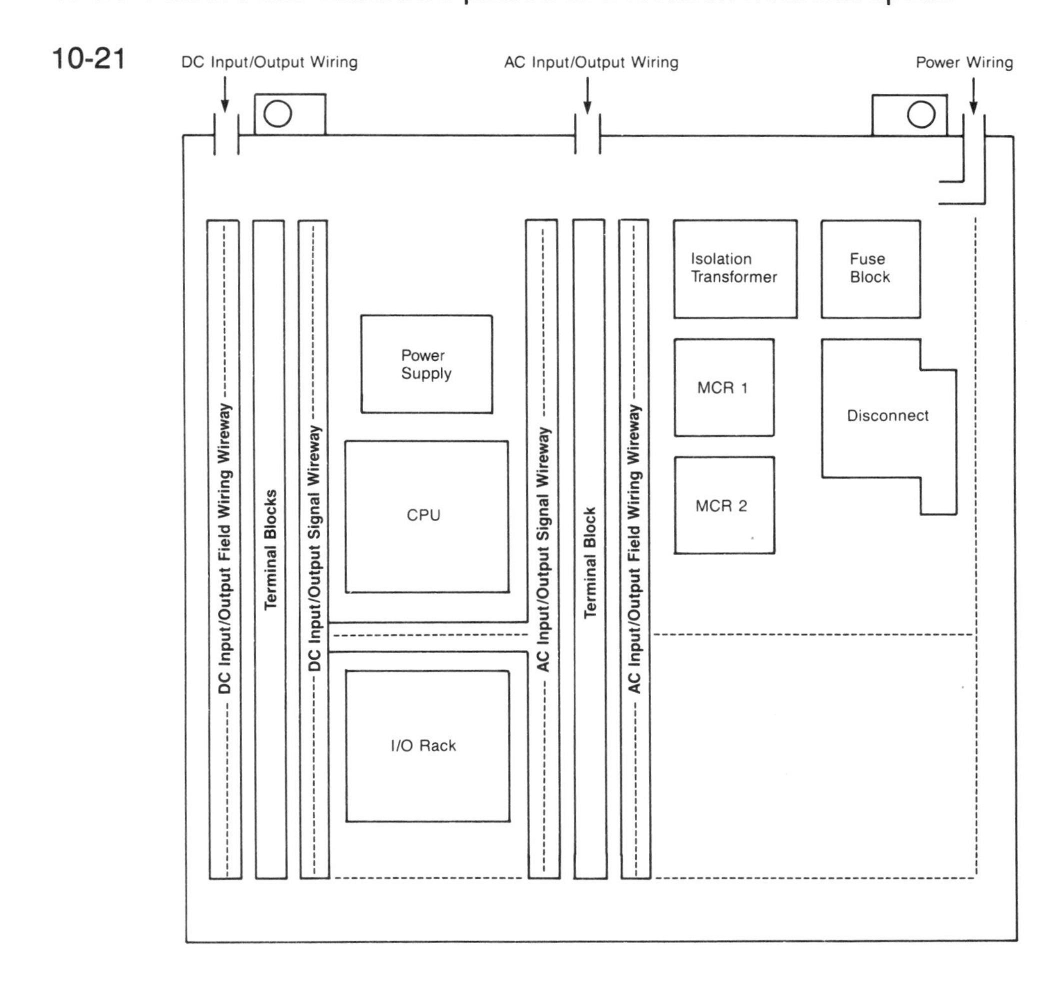

10-22 Cross talk

10-23 The duct and wiring layout defines the physical location of wireways and routing of field I/O signals, power, and PC interconnections within the enclosure.

10-24 True

10-25 True

10-26 Separate

10-27 True

10-28 Right angles

10-29 Grounding

10-30 Permanent (not soldered)

10-31 True

10-32 Paint or non-conductive materials should be scraped away to provide good ground connection.

10-33 True

10-34 Isolation transformers are desirable in cases in which heavy equipment is likely to introduce noise onto the AC line.

10-35 Relays, motor starters, motors, solenoids, and auxiliary contactors.

10-36 True

10-37 Should not

10-38 False. Emergency stops should be used when necessary to maintain the safety of the control system.

10-39 a- Electromechanical MCRs

10-40 Outrush is a condition that occurs when output triacs are turned off by throwing the power disconnect, thus causing the energy stored in an inductive load to seek the nearest path to ground, which is often through the triacs. To correct this problem, a capacitor may be placed across the disconnect (.47 uF for 120 VAC, .22 uF for 240 VAC).

10-41 c- 60% of inputs and 30% of outputs are ON at one time

10-42 Constant voltage transformer, soft AC lines

10-43 b- Every time there is a change

10-44 d- None of the above

10-45 False. The input power and the AC and DC wire bundles should be kept separate.

10-46 d- All of the above

10-47 True

10-48 True

10-49 Special wiring considerations may be necessary for leaky inputs inductive load (suppression), output fusing, and the proper wiring of low-level and analog signals (may require shielded cable).

10-50 OFF

10-51 False

10-52 Bleeding resistor

10-53 c- Metal oxide varistor

10-54 True

10-55 a- Small AC load suppression
c- DC load suppression
b- Large AC load suppression

10-56 True

10-57 d- All of the above

10-58 True

10-59 By manually activating the device

10-60 False

10-61 Disconnected

10-62 Dummy rung

10-63 d- All of the above

10-64 Changes should be made immediately.

10-65 During the scheduled maintenance of the machine.

10-66 When fans are used, filters should be changed periodically, depending on the environment where the PC enclosure is located.

10-67 Any build-up of dust or dirt can obstruct the heat dissipation of components in the system.

10-68 True. This check would avoid any major problems in trying to find a fault due to a loose terminal connection.

10-69 d- All of the above

10-70 d- a and c. A good practice is to have about 10% of input and output modules plus a power supply and one of each of the main boards that form the PC system.

10-71 d- All of the above

10-72 d- All of the above

10-73 The first check should be the input power and/or logic indicators.

10-74 1- f and g, the power and logic indicators
2- h and i, check for the connection between the common (input module) and the line common (L2)
3- e, check for a good terminal connection at the input module
4- a, check to see if the device is connected to the AC line (L1)
5- b, check for power at the device itself
6- c and d, by manually activating the input device, check for proper operation; the voltage at b should be present at d when the device is activated

This procedure applies if the input module is in the master or CPU rack; otherwise, the first thing to check is the I/O rack or communications cable to insure that the cable is present.

10-75 b- Terminal connector at the input card may be faulty. This problem seldom occurs, but when it does, it is very difficult to detect.

10-76 b- Place a voltmeter across the load and measure the voltage.

10-77 c- Isolate the problem to the module or to the field wiring.

UNIT
--]11[--

ANSWERS

11-1 Several PCs and/or host processors

11-2 Node

11-3 Capable of supporting real-time control
High data integrity
High noise immunity
High reliability in harsh environments
Suitable for large installations

11-4 Business networks do not require as much noise immunity as industrial networks because they are intended to be used in an office environment. The access-time requirements in a business network are less stringent; a business workstation can easily wait a few seconds for information, but a machine being controlled by a PC may need information within milliseconds.

11-5 Communications between PCs were accomplished by connecting an output module of one PC to an input module of another PC. The amount of data to be transferred and the speed would determine how many wires would be used and the method (parallel or series transmission).

11-6 Centralized data acquisition and distributed control

11-7 When a single PC is used for data collection and control in a large process, the amount of memory consumed is extensive and tends to slow down the scan time and complicate the control logic program.

11-8 True. The performance of each system is independent of others, thus having a fast scan time; the reliability increases since one PC can be down and the remainder of the system will continue operation.

11-9 True. Topology defines the geometry configuration or connection of each node in the network.

11-10 d- All of the above

11-11 c- Tree

11-12 False. There is no tree topology, therefore, there are no differences.

11-13 True

11-14 c- Failure of a central node will bring down the system.

11-15

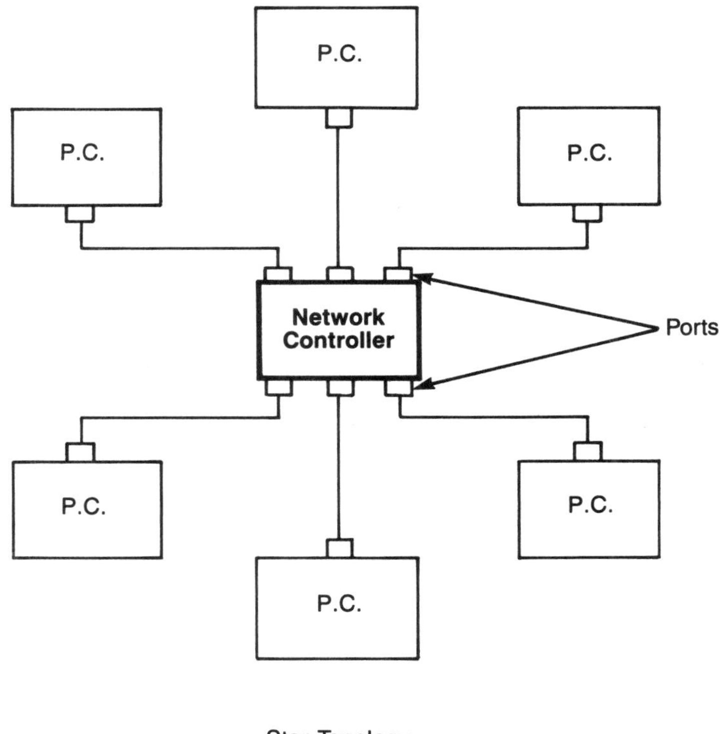

Star Topology

11-16 Multidrop

11-17 b- Each station has equal independent control.
c- Reconfiguration of networks is easy. High speed throughout does not depend on topology, but rather on the access method and the speed of the network.

11-18 True. A break in the trunkline could affect many nodes.

11-19 False. Since a break can affect at least one node, the communication of the network is deteriorated.

11-20 True

11-21 In a master-slave configuration, the master sends data to the slaves; if data is needed from a slave, the master polls the slave (address) and waits for a response. No communication takes place unless it is initiated by the master.

11-22 b- Failure of a node brings down the system.

11-23 True

11-24 c- It requires twice as much wiring.

11-25 b- Access method

11-26 Common bus topology

11-27 Polling

11-28 The master

11-29 c- Carrier Sense Multiple Access with Collision Detection

11-30 In the collision detection access method, each node that has a message to transmit waits until there is no traffic on the network and then transmits. Once a node is transmitting, the collision-detection circuitry is checking for the presence of another transmitter. If two nodes transmit at the same time, a collision occurs and is detected by the circuitry; the transmitter is then disabled, and the node waits a variable amount of time before retrying.

11-31 c- There are not many nodes in the network. The more nodes, the greater the possibility of multiple transmissions and, therefore, of collision.

11-32 b- Throughput drops off and access time increases.

11-33 False

11-34 c- Contention among stations that are trying to gain access to the highway

11-35 Token

11-36 f- Relinquish the highway
d- Next designated node

11-37 False. The token is passed in sequential manner.

11-38 In a common bus network, each station is identified by an address.

11-39 c- Many nodes and/or stringent response time

11-40 True

11-41 b- Point-to-point applications

11-42 a- Reflection of transmission
b- Nonuniformity of the cable impedance
c- Problem with distance

11-43 Broadband coaxial cable

11-44 False

11-45 True

11-46 b- More cost
c- Small and light weight
d- Totally immune to electrical interference

11-47 True

11-48 The maximum length of the main trunkline cable, and the maximum length of the "drop" or length of the device and the main line

11-49 True. Specific recommendations are given for the coaxial cable to be used depending on the distance.

11-50 b- The maximum response time

11-51 Gateway

11-52 True

11-53 a- Information stored in the executive

11-54 A protocol is a set of rules that must be followed if two or more devices are to communicate with each other.

11-55 A protocol generally defines the following information:

Communication line errors
Flow control to keep buffers from overflowing
Access by multiple devices.
Failure of detection
Data translation
Interpretation of messages

11-56 e- Layer 7. Application
a- Layer 6. Presentation
g- Layer 5. Session
c- Layer 4. Transport
d- Layer 3. Network
b- Layer 2. Data link
f- Layer 1. Physical

11-57 True

11-58 c- Common bus topology
f- Token pass access method

11-59 j- Broadband
f- Baseband
i- Local Area Networks
b- Polling
d- Twisted pair cable
h- ISO/OSI
k- Coaxial cable
e Gateway
a Node
g Token pass
c Fiber optics

UNIT

--]12[--

ANSWERS

12-1 Programmable controllers are used in the glass industry in the process and forming of glass, annealing Lehr control, cullet weighing, batching, pelletizing, and packaging and in the material handling process.

12-2 In the chemical and petrochemical industry, programmable controllers are used to control the batch process, the weighing and mixing of chemicals, pipeline pump control, and in off-shore drilling.

12-3 A PC system can control all the sequential operations, alarms, and safety logic necessary to load and circulate parts from the production area to the finished goods warehouse, as well as the sorting of products to correct conveyor lanes. The system can keep track of the produced goods and send the information to a central host that keeps the amount of product in warehouse inventory.

12-4 A programmable control system can not only control the flow of materials in an assembly line but also monitor engine test and record data from sensors such as water temperature, oil temperature and pressure, RPMs, manifold pressure, and other vital measurements.

12-5 A PC with a DC power supply could be used to monitor the AC line by connecting one or more AC input modules to the power line. If a failure in the line occurs, the input modules would be de-energized, and the system could go into a generator start sequence (after a small delay). The application software would be controlling the sequence of events that must take place in starting the auxiliary generator and other machinery.

12-6 Programmable controller systems are used in modernization projects primarily because PCs bring more reliability into a system, and the time and money of implementation are much less than a conventional electromechanical control system.

12-7 A PC-based system is useful in a batching application because it can not only measure signals from load cells to determine the proper weight for a batch, but also adjust parameters of ingredients based on recipes or formulas stored in the system. The programmable controller can also control feeders and valves for infeed and outfeed from weight hoppers, shut-off gates, and other equipment.

12-8 OEMs use programmable controllers not just because of price justifications and turnaround of a machine, but also the ability to change parameters in similar machines with different control schemes. The improvement in reliability of the machine benefits the OEM as well as its customer, the end user.

12-9 Applications in which the control task can be separated are more suitable for distributed control because the control application program is simplified for each of the stations. Furthermore, each station is independent of other PCs while interlocking, if required, can be done through the data highway. If one station is down, production continues, thus minimizing total down time.

12-10 Some applications, most of them modernizations, can justify the use of a PC to replace only timers because the cost associated with a small controller could be justified with as few as ten electromechanical timers. The adjustment of the timers can be precise and easily changed if necessary. More control can also be incorporated in the future if required.

UNIT

--]13[--

ANSWERS

13-1 Relay replacers

13-2 f- c and d

13-3 This curve describes product ranges for programmable controllers, which are subdivided into four major areas (1, 2, 3, and 4), with overlapping boundaries (a, b, and c).

The basis for segmentation of the areas includes the number of possible inputs and outputs a system can accommodate as well as the amounts of memory and software capabilities.

The overlapping areas indicate products that can fall in one of the major areas (a lower one), but have features that are found in the next higher range.

13-4 The digital I/O count

13-5 False. As the I/O count increases, so does the complexity of the system, including the amount of memory.

13-6 True

13-7 d- 32 or less I/O

13-8 False

13-9 b-2

13-10 b- Enhancements to standard features of products in segment 2

13-11 a- True
b- False
c- True
d- False
e- True

13-12 Large PCs include all the software capabilities of a medium-size PC; some of the extra features are double precision arithmetic, more block transfers, PID capabilities, host computer communication modules, more than one RS-232C communication port, more memory, and subroutine capabilities.

13-13 d- All of the above

13-14 False. Future needs must also be considered.

13-15 c- Documentation

13-16 The amount of I/O, whether digital and/or analog

13-17 True

13-18 Fuses
Transient surge protection
Isolation between power and logic
Indicators
Cost per point

13-19 b- The accessibility

13-20 c- Isolated commons

13-21 c- Bipolar
d- Special analog input
b- 4 to 20 mA
a- Special I/O
f- PID module
e- Analog I/O

13-22 Reduce

13-23

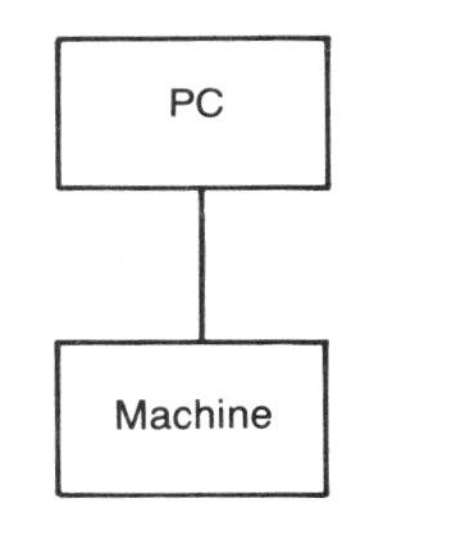

a) Individual Machine Control

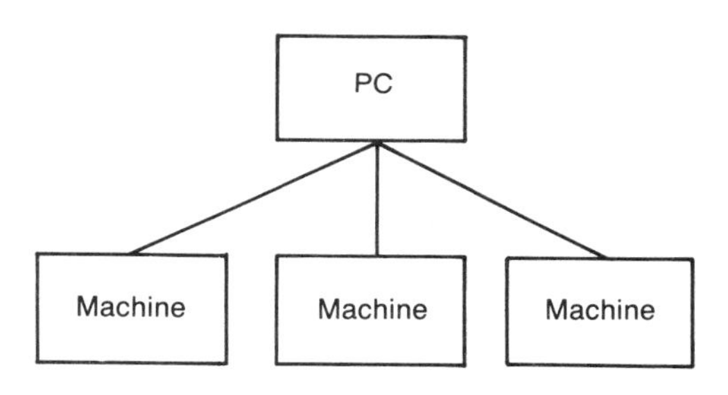

b) Centralized Control

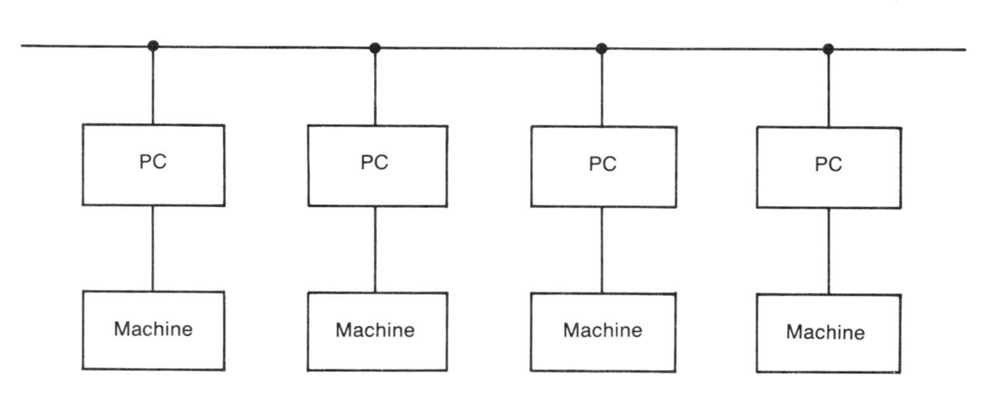

c) Distributed Control

13-24 True

13-25 b- Used in more than one machine

13-26 True

13-27 b- It avoids problems of decentralizing the control task into smaller ones.

13-28 Redundant

13-29 c- Two PCs in communication

13-30 Data highway (or Local Area Network)

13-31 False. Different highways may have completely different protocols, ways of accessing nodes, and speed.

13-32 The type and amount of memory

13-33 True

13-34 c- Inputs and outputs
d- Complexity of control program

13-35 True

13-36 Programming device

13-37 True. Because the CPU determines the type and number of peripherals that can be interfaced and supported as well as the method of interfacing.

13-38 False

13-39 b- Operating parameters
c- Packaging

13-40 False. Technical support of a programmable controller product is important regardless of how capable the in-house engineering group may be.

13-41 b- Before the purchase of equipment

13-42 True

13-43 Burn-in is a procedure by which products are cycled through high and low temperature for a period of time (minimum of 48 hours). The purpose of burn-in is to take the infant mortality out of the hardware, so that products that pass the test will most likely not fail in the field under normal circumstances.

13-44 c- Standardization

13-45 d- All of the above

13-46 a- Step 1
c- Step 2
e- Step 3
g- Step 4
b- Step 5
d- Step 6
f- Step 7
i- Step 8
j- Step 9
h- Step 10
k- Step 11

DO NOT RETURN IN JIFFY BAG